CHECKP✓INT
AND BEYOND

Complete Physics for Cambridge Secondary 1

Helen Reynolds

OXFORD
UNIVERSITY PRESS

Contents

Stage 9

How to use your Student Book

Welcome to your **Complete Physics for Cambridge Secondary 1** Student Book. This book has been written to help you study Physics at all three stages of Cambridge Secondary 1.

Most of the pages in this book work like this:

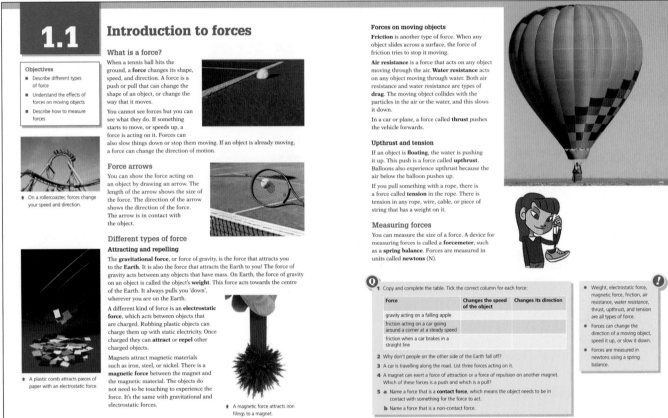

- Every page starts with the learning objectives for the lesson. The learning objectives match the Cambridge Secondary 1 Science curriculum framework.

- New vocabulary is marked in bold. You can check the meaning of these words in the glossary at the back of the book.

- At the end of each page there are questions to test that you understand what you have learned.

- The key points to remember from the page are also summarised here.

These pages cover the Physics topics in the Cambridge Secondary 1 Science curriculum framework. In addition, in every chapter there are also pages that help you think like a scientist, prepare for the next level, and test your knowledge. Find out more on the next page.

Scientific enquiry

These pages help you to practise the skills that you need to be a good scientist. They cover all the scientific enquiry learning objectives from the curriculum framework.

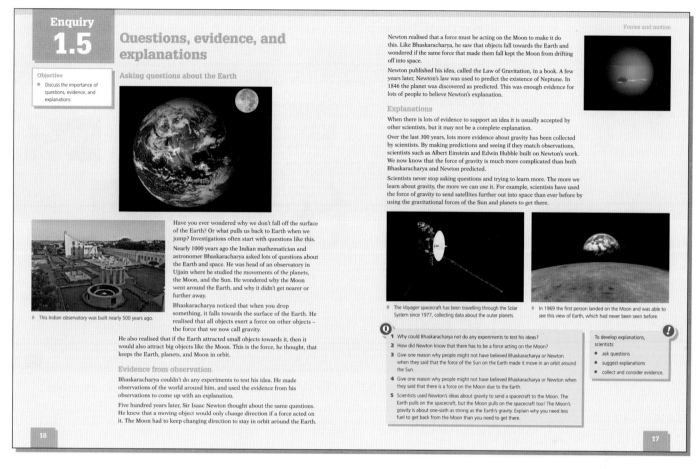

You will learn how to:

- consider ideas
- plan investigations and experiments
- record and analyse data
- evaluate evidence to draw scientific conclusions.

You will also learn how scientists throughout history and from around the globe created theories, carried out research, and drew conclusions about the world around them.

Extension

Throughout this book there are lots of opportunities to learn even more about physics beyond the Cambridge Secondary 1 Science curriculum framework. These topics are called *Extension* because they extend and develop your science skills even further.

You can tell when a topic is extension because it is marked with a dashed line, like the one on the left. Or when the page has a purple background, like below.

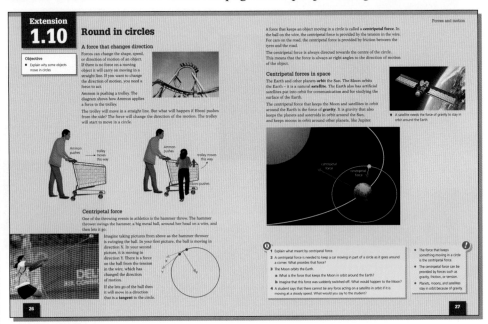

Extension topics will not be in your Cambridge Checkpoint test, but they will help you prepare for moving onto the next stage of the curriculum and eventually for Cambridge IGCSE® Physics.

Review

At the end of every chapter and every stage there are review questions.

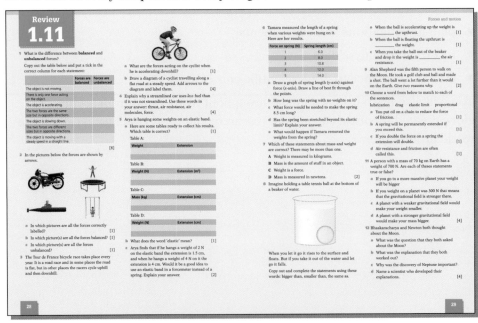

These questions are written in the style of the Cambridge Checkpoint test. They are there to help you review what you have learned in that chapter or stage.

Reference

At the back of this book there are reference pages. These pages will be useful throughout every stage of Cambridge Secondary 1 Science.

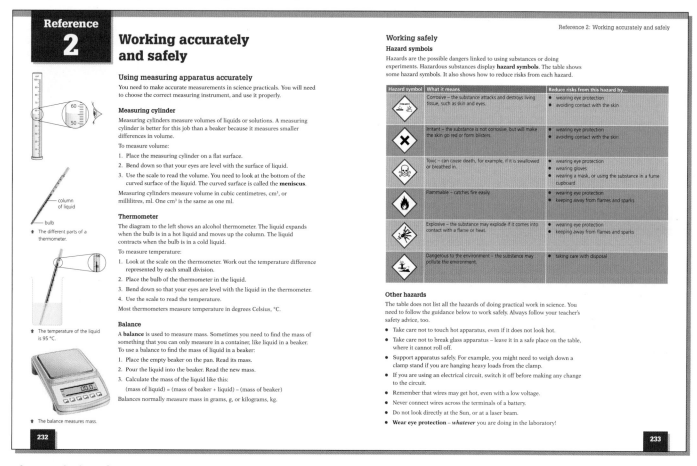

They include information on:

- how to choose suitable apparatus
- how to work accurately and safely
- how to record, display, and analyse results
- how to use ammeters and voltmeters.

1.1

Introduction to forces

Objectives

- Describe different types of force
- Understand the effects of forces on moving objects
- Describe how to measure forces

What is a force?

When a tennis ball hits the ground, a **force** changes its shape, speed, and direction. A force is a push or pull that can change the shape of an object, or change the way that it moves.

You cannot see forces but you can see what they do. If something starts to move, or speeds up, a force is acting on it. Forces can also slow things down or stop them moving. If an object is already moving, a force can change the direction of motion.

⬆ On a rollercoaster, forces change your speed and direction.

Force arrows

You can show the force acting on an object by drawing an arrow. The length of the arrow shows the size of the force. The direction of the arrow shows the direction of the force. The arrow is in contact with the object.

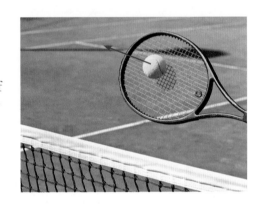

Different types of force

Attracting and repelling

The **gravitational force**, or force of gravity, is the force that attracts you to the **Earth**. It is also the force that attracts the Earth to you! The force of gravity acts between any objects that have mass. On Earth, the force of gravity on an object is called the object's **weight**. This force acts towards the centre of the Earth. It always pulls you 'down', wherever you are on the Earth.

A different kind of force is an **electrostatic force**, which acts between objects that are charged. Rubbing plastic objects can charge them up with static electricity. Once charged they can **attract** or **repel** other charged objects.

Magnets attract magnetic materials such as iron, steel, or nickel. There is a **magnetic force** between the magnet and the magnetic material. The objects do not need to be touching to experience the force. It's the same with gravitational and electrostatic forces.

⬆ A magnetic force attracts iron filings to a magnet.

⬆ A plastic comb attracts pieces of paper with an electrostatic force.

Forces on moving objects

Friction is another type of force. When any object slides across a surface, the force of friction tries to stop it moving.

Air resistance is a force that acts on any object moving through the air. **Water resistance** acts on any object moving through water. Both air resistance and water resistance are types of **drag**. The moving object collides with the particles in the air or the water, and this slows it down.

In a car or plane, a force called **thrust** pushes the vehicle forwards.

Upthrust and tension

If an object is **floating**, the water is pushing it up. This push is a force called **upthrust**. Balloons also experience upthrust because the air below the balloon pushes up.

If you pull something with a rope, there is a force called **tension** in the rope. There is tension in any rope, wire, cable, or piece of string that has a weight on it.

Measuring forces

You can measure the size of a force. A device for measuring forces is called a **forcemeter**, such as a **spring balance**. Forces are measured in units called **newtons** (N).

1 Copy and complete the table. Tick the correct column for each force:

Force	Changes the speed of the object	Changes its direction
gravity acting on a falling apple		
friction acting on a car going around a corner at a steady speed		
friction when a car brakes in a straight line		

2 Why don't people on the other side of the Earth fall off?

3 A car is travelling along the road. List three forces acting on it.

4 A magnet can exert a force of attraction or a force of repulsion on another magnet. Which of these forces is a push and which is a pull?

5 a Name a force that is a **contact force**, which means the object needs to be in contact with something for the force to act.

   b Name a force that is a non-contact force.

- Weight, electrostatic force, magnetic force, friction, air resistance, water resistance, thrust, upthrust, and tension are all types of force.

- Forces can change the direction of a moving object, speed it up, or slow it down.

- Forces are measured in newtons using a spring balance.

Balanced forces

Balanced and unbalanced forces

You can use forces to explain why an object is moving in the way that it is, or why it is not moving. The force of gravity, or weight, is always acting on the diver. Why is she moving in only one of the pictures?

If the forces on an object are the same size but in opposite directions, then they will cancel out. The forces are **balanced**. It's a bit like a tug-of-war when the teams are equal. The object behaves as if there is *no* force acting on it. When the diver is floating, her weight is balanced by the upthrust of the water.

If the forces acting on an object are balanced, then its motion will not change:

- If it is not moving it will stay still.
- If it *is* moving it will keep moving at a steady speed.

If the forces on an object are *not* equal and opposite, then they are **unbalanced**. If the forces are unbalanced then the motion of the object *will* change:

- If it is not moving it will start moving.
- If it is moving it will speed up (**accelerate**), slow down (**decelerate**), or change direction.

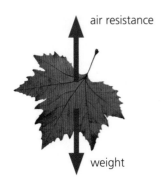

air resistance

weight

⬆ The weight of the leaf and the air resistance are balanced. The leaf falls with a steady speed.

➡ The forces are unbalanced. The cyclist accelerates.

friction and air resistance

thrust

Lots of people think if something is moving, like a football rolling along the ground, that there is a force acting on it. This is not the case. The football accelerates while it is in contact with the player's boot. Once it is no longer touching his boot the only forces acting on it come from the things that it is in contact with – the ground and the air. Friction and air resistance will slow it down. If we could remove both of these forces, then the ball would carry on travelling at a steady speed forever. You will learn more about a scientist called Galileo who had this idea on pages 90–91.

Lots of forces

Sometimes more than one pair of forces is acting on an object. Look at the boat. The weight and the upthrust are balanced. The boat does not move up or down.

The thrust of the engine is balanced by the air and water resistance. The boat moves forwards with a steady speed.

You know that if an object is stationary or moving at a steady speed, then the forces on it are balanced. If it is speeding up or slowing down, then the forces acting on it will be unbalanced.

Resultant forces

If you know the size of each force acting on an object, you can work out the **resultant force**. If the arrows are in the same direction you add the forces. If they are in opposite directions you take away one force from the other.

2 N   6 N

⬆ The resultant force is 4 N to the right.

10 N   2 N

⬆ The resultant force is 8 N to the left.

2 N   2 N

⬆ The resultant force is zero.

- If the resultant force is zero the forces are **balanced**.
- If the resultant force is not zero the forces are **unbalanced**.

Q

1 A cricketer drops a ball and it speeds up. Are the forces on it balanced or unbalanced?

2 A boy pushes a toy car along the floor at a steady speed. Are the forces on it balanced or unbalanced?

3 Arzu is throwing a ball up in the air and catching it. Copy and complete the sentences using the words 'balanced' or 'unbalanced'.

  a When Arzu is holding the ball and it is not moving, the forces on the ball are … .

  b When the ball is moving upwards and slowing down, the forces on it are … .

  c When the ball is moving downwards and speeding up, the forces on it are … .

4 **Extension:** Vil says that if a car is moving there must be a resultant force acting on it. Alom says there doesn't need to be a resultant force acting on it. Who do you agree with, and why?

!

- Forces are balanced if they cancel each other out. They are unbalanced if they do not cancel out.

- An object stays still or moves at a steady speed if the forces on it are balanced.

- Unbalanced forces change the speed or direction of a moving object.

1.3

Friction

What is friction?

Friction is a force that slows down moving objects.

When a child goes down a slide, there is a force of friction between them and the slide. This is because all surfaces, even surfaces that feel very smooth, are uneven. A metal slide looks smooth, but under the microscope you can see how uneven it is.

You need to push things to get them to move. It is the uneven surfaces that produce the force of friction that you have to overcome.

Objectives

- Describe the effect of friction on moving objects
- Understand how to reduce friction
- Describe how friction can be useful

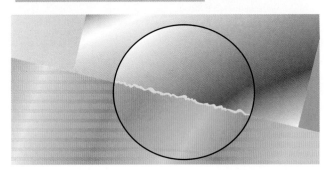

⬆ Even smooth surfaces like shiny metal are rough if you look at them under a microscope.

Reducing friction

People often try and reduce friction, for example skiers wax the bottoms of their skis to reduce friction and make them go faster.

Cyclists put oil on the axles of their bicycle to reduce friction. This is called **lubrication**. A layer of oil between two surfaces makes it easier for the surfaces to slide over each other. It reduces the force of friction.

There are ball-bearings inside the wheels of a skateboard. They roll over each other as the wheel turns, and reduce friction. This means that the surfaces are not worn away as quickly.

Friction can be useful

Friction is not always a bad thing. When you walk friction is useful. Friction acts between your feet and the ground, making it possible for you to move. Vehicles also need the force of friction between the tyres and the road to make them move. In icy conditions friction is reduced and the wheels skid because there is not enough friction for the tyres to grip the road. When there is not much friction, you realise how important friction is!

Bicycles and cars also need friction to stop. The force of friction acts between their brakes and the wheels. When surfaces rub against each other like this they get warm, just like when you rub your hands together. The temperature of brake pads can get very high if the car is moving fast and it brakes hard. Sometimes they get so hot that they glow. The friction between brake pads and the wheels wears them away and they have to be replaced.

Q

1 What causes the force of friction between two surfaces that are sliding over each other?

2 Explain how using oil for lubrication reduces friction.

3 a There is a thin layer of water between the blade and the ice. Suggest how this affects friction and helps the skater.

   b Suggest why ice skates have a jagged edge at the front.

4 Tyres have treads to remove water from between the tyre and the road. Why is it important to remove the water?

- The force of friction slows things down.
- You can reduce friction by lubrication or by using ball-bearings.
- Friction is useful for grip, to start moving, or for braking.

1.4 Gravity

Objectives

- Explain the link between gravity, mass, and weight
- Describe how your weight can be different on different planets

The force of gravity

There is a force between all objects called the force of **gravity**. This force is pulling you down towards the centre of the Earth. But forces come in pairs – you are also pulling the Earth with a force of gravity.

The size of the force of gravity between two objects depends on:

- the mass of the two objects
- how far apart they are.

Weight and mass

The force of the Earth's gravity on an object is called its **weight**. When you use scales or a forcemeter you are measuring the force of gravity.

⬆ There is a force of gravity between you and the Earth

Weight is a force so it is measured in **newtons** (N). **Mass** is related to the amount of **matter** in an object and it is measured in **kilograms** (kg). Another way of thinking about mass is to do with motion. Something with a small mass will accelerate faster when you apply a force to it than something with a big mass.

SUGAR

It's easy to confuse weight and mass. Your weight is a force and should be measured in newtons, but often scales measure your 'weight' in kilograms! Scientifically this is wrong.

Your weight can change

The force of gravity is different on every planet. The more massive the planet, the larger the force of gravity. If you could visit different planets your weight would change, but your mass would stay the same.

On the Moon there is no atmosphere (air) and astronauts have to wear spacesuits to be able to breathe. Some people think this means there is no gravity there. This is not true. There is a force of gravity on the Moon, but it is about one-sixth as strong as on Earth.

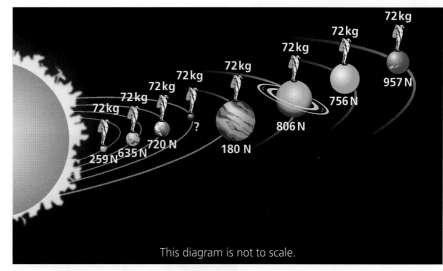

This diagram is not to scale.

⬆ The weight of the astronaut changes, but his mass stays the same.

Gravitational field strength

There is a **gravitational field** around all objects. A gravitational field is a region where a mass experiences a force.

On Earth the **gravitational field strength** is 10 newtons per kilogram (N/kg). To work out the weight of an object use this equation:

Weight = mass × gravitational field strength
  (N)   (kg)   (N/kg)

When people say that gravity is different on different planets, they mean that the gravitational field strength is different.

In physics the idea of a 'field' is very important. Fields are linked to forces. There are other types of field, such as magnetic fields (see page 138).

Q

1 Explain the difference between weight and mass.

2 A baby has a mass of 4 kg. What is its weight on Earth?

3 Is the mass of the astronaut on Mars bigger, smaller, or the same compared with on Earth? Explain your answer.

4 A student says that objects get pulled down because the Earth is like a big magnet.

   a How is the gravitational force like a magnetic force?

   b How is the gravitational force *not* like a magnetic force?

5 **Extension:** the gravitational field strength on Mars is 4 N/kg. Calculate the weight of the astronaut on Mars.

!

- The force of gravity on an object on Earth is called its weight. It depends on the mass.

- Weight is a force, measured in newtons.

- Mass is the amount of matter, measured in kilograms.

- The force of gravity is different on different planets or on the Moon.

Questions, evidence, and explanations

Asking questions about the Earth

Have you ever wondered why we don't fall off the surface of the Earth? Or what pulls us back to Earth when we jump? Investigations often start with questions like this.

Nearly 1000 years ago the Indian mathematician and astronomer Bhaskaracharya asked lots of questions about the Earth and space. He was head of an observatory in Ujjain where he studied the movements of the planets, the Moon, and the Sun. He wondered why the Moon went around the Earth, and why it didn't get nearer or further away.

↑ This Indian observatory was built nearly 500 years ago.

Bhaskaracharya noticed that when you drop something, it falls towards the surface of the Earth. He realised that all objects exert a force on other objects – the force that we now call gravity.

He also realised that if the Earth attracted small objects towards it, then it would also attract big objects like the Moon. This is the force, he thought, that keeps the Earth, planets, and Moon in orbit.

Evidence from observation

Bhaskaracharya couldn't do any experiments to test his idea. He made observations of the world around him, and used the evidence from his observations to come up with an explanation.

Five hundred years later, Sir Isaac Newton thought about the same questions. He knew that a moving object would only change direction if a force acted on it. The Moon had to keep changing direction to stay in orbit around the Earth.

Newton realised that a force must be acting on the Moon to make it do this. Like Bhaskaracharya, he saw that objects fall towards the Earth and wondered if the same force that made them fall kept the Moon from drifting off into space.

Newton published his idea, called the Law of Gravitation, in a book. A few years later, Newton's law was used to predict the existence of Neptune. In 1846 the planet was discovered as predicted. This was enough evidence for lots of people to believe Newton's explanation.

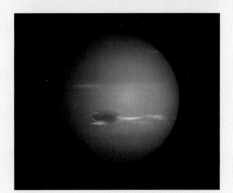

Explanations

When there is lots of evidence to support an idea it is usually accepted by other scientists, but it may not be a complete explanation.

Over the last 300 years, lots more evidence about gravity has been collected by scientists. By making predictions and seeing if they match observations, scientists such as Albert Einstein and Edwin Hubble built on Newton's work. We now know that the force of gravity is much more complicated than both Bhaskaracharya and Newton predicted.

Scientists never stop asking questions and trying to learn more. The more we learn about gravity, the more we can use it. For example, scientists have used the force of gravity to send satellites further out into space than ever before by using the gravitational forces of the Sun and planets to get there.

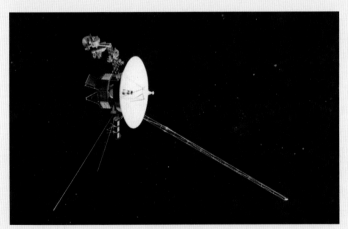

⬆ The *Voyager* spacecraft has been travelling through the Solar System since 1977, collecting data about the outer planets.

⬆ In 1969 the first person landed on the Moon and was able to see this view of Earth, which had never been seen before.

Q

1 Why could Bhaskaracharya not do any experiments to test his ideas?

2 How did Newton know that there has to be a force acting on the Moon?

3 Give one reason why people might not have believed Bhaskaracharya or Newton when they said that the force of the Sun on the Earth made it move in an orbit around the Sun.

4 Give one reason why people might not have believed Bhaskaracharya or Newton when they said that there is a force on the Moon due to the Earth.

5 Scientists used Newton's ideas about gravity to send a spacecraft to the Moon. The Earth pulls on the spacecraft, but the Moon pulls on the spacecraft too! The Moon's gravity is about one-sixth as strong as the Earth's gravity. Explain why you need less fuel to get back from the Moon than you need to get there.

!

To develop explanations, scientists:

● ask questions

● suggest explanations

● collect and consider evidence.

1.6

Air resistance

What is air resistance?

⬆ The air can produce a force that changes your face!

When an object moves through air, there is a force on it called **air resistance**. Most of the time you don't really notice it. If the air is moving very fast, or you are moving very fast through the air, then you notice the effect.

An object moving through the air collides with the particles in the air and is effectively pushing the air out of the way. These collisions with air particles provide the resistance.

Air resistance is also affected by the *speed* of the object moving through the air. Objects moving with a higher speed will push more air out of the way and experience more air resistance.

Reducing air resistance

Air resistance can be a problem. It is a form of friction that slows things down. Air resistance is less if the area in contact with the air is reduced. **Streamlining** reduces air resistance by changing the flow of air over a car or plane. Scientists use wind tunnels to experiment with the shape of vehicles and find the best shape.

⬆ Smoke shows the path of air in a wind tunnel.

Cyclists pull in their arms and crouch forward to reduce the area in contact with the air. They make themselves more streamlined by using special helmets.

Using air resistance

Air resistance can be very useful for slowing things down. A parachute increases the area that is in contact with the air, and therefore increases the air resistance.

When a parachutist jumps out of a plane she accelerates because of the force of gravity acting on her. As she speeds up the air resistance increases until it balances her weight. She then falls with a steady speed called the **terminal velocity**.

⬆ Rocket cars use parachutes for braking.

When she opens her parachute the air resistance suddenly increases. This slows her down. As she slows down the air resistance decreases until she reaches a new, slower, terminal velocity and can land safely.

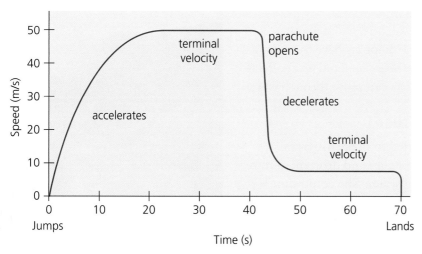

⬆ This graph shows how the speed of a parachutist changes as she falls.

Do heavier things fall faster?

When we see things fall on Earth, like a feather or a hammer, the heavier object (the hammer) falls faster. This is because of air resistance. Air resistance affects the motion of the feather much more than the motion of the hammer. If you remove the air, or do the experiment somewhere where there is no air like the Moon, the hammer and feather hit the ground at the *same* time. If there is no air resistance all objects fall at the same rate.

Q

1 Car manufacturers put cars in wind tunnels to help them to design streamlined cars.

  a Explain what is meant by 'streamlined'.

  b Which would experience more air resistance – a streamlined car travelling slowly or a lorry travelling fast? Explain your answer.

2 Copy and complete these sentences about the forces on a skydiver jumping out of a plane using the words 'balanced' or 'unbalanced'.

  When the skydiver jumps out of the plane the forces are _____. When she reaches a steady speed of about 50 m/s the forces are _____. The parachute then opens and this makes the forces _____. She then reaches a lower terminal velocity of about 5 m/s and the forces are _____. Finally the skydiver lands and stands on the ground. The forces are _____.

3 Explain why a tennis ball and a cricket ball dropped together will hit the ground at about the same time, even though the cricket ball is heavier.

!

- Air resistance depends on the speed of the object and its area in contact with the air.

- The shape of streamlined objects reduces air or water resistance.

- Objects fall at terminal velocity when their weight is balanced by air resistance.

- A parachute reduces the terminal velocity so a skydiver can land safely.

Planning investigations

Objective

■ Understand how to plan an investigation to test an idea in science

How does the mass affect the time to hit the bottom?

How does the shape affect the time to hit the bottom?

Asking questions

Kasini was watching a film about dolphins. The dolphins have to swim fast to catch fish. She wondered what affects how fast things can move through water.

Ideas to test

Kasini decided to make different objects out of clay and drop them into a cylinder of water. She could time how long they took to hit the bottom. These are some of the ideas that she thought about testing.

Investigations are ways of obtaining **evidence**. Before Kasini could write down a plan for her **investigation**, she had to decide exactly which question she wanted to answer. This is what she wrote:

> I have decided to investigate how the shape of the object affects how long it takes to fall through water.

In an investigation something that can be changed is called a **variable**. This is the list of possible variables that Kasini thought of:

● the shape of the clay

● the mass of clay

● the volume of water in the cylinder

● the temperature of the water.

The one variable that she decided to change was the shape of the clay. It is very important to change only one variable at a time. She decided that she would use the same mass of clay and the same volume of water at the same temperature each time.

Making a prediction

Kasini had learned that engineers design cars and aeroplanes to be as streamlined as possible to reduce drag. She knew that air resistance is a form of drag, and so is water resistance. She used this information to make a prediction.

> I predict that the cone shape will reach the bottom in the shortest time. This is because there will be less water resistance because a cone is a streamlined shape.

Making a plan and choosing equipment

This is Kasini's plan.

> I am going to investigate how long it takes different shapes of clay to reach the bottom of the cylinder of water. I will make different shapes from the same amount of clay. These are the shapes I have chosen: cone, cube, sphere, cylinder, cuboid.
>
> I will time how long it takes for the shape to hit the bottom with a stopwatch.
>
> This is a list of my equipment:
> - a large measuring cylinder
> - modelling clay
> - a stopwatch
> - a balance
> - a measuring jug.
>
> I will write my results in a table.

Making improvements

Kasini completed her investigation and wrote down her results.

She discussed her investigation with Nadia. Nadia asked if there were any problems with the investigation. Kasini said that it was difficult to see exactly when to start and stop the stopwatch because all the shapes moved quickly through the water.

Shape	Time
cone	0.58
cube	0.65
sphere	0.61
cylinder	0.75
cuboid	0.68

1 Copy and complete this table to explain why Kasini needed each piece of equipment.

Equipment	Why Kasini needed it
a large measuring cylinder	
modelling clay	
a stopwatch	
a balance	
a measuring jug	

2 Why is it important to change only one variable at a time?

3 Sometimes it is hard to see when to start and stop the stopwatch. How could the plan be improved to give better results?

4 What has Kasini missed out of her results table?

5 Nadia says that two variables that Kasini should have in her list are the type of stop clock and the type of clay. Do you agree? Explain your answer.

- An idea can be tested by carrying out an investigation.
- Variables are things that you can control, change, or observe.
- You can use scientific knowledge to make predictions.

Tension and upthrust

Objectives

- Describe what happens when you stretch a spring
- Explain what is meant by tension
- Explain the elastic limit
- Explain why things float or sink

Under tension

Ropes for rock climbing are designed to **stretch** a certain amount, just in case the climber falls. If the rope did not stretch then he could be injured. Ropes, cords, and springs all stretch when you pull them.

There is force in the rope that balances the climber's weight. The force in a rope that is being stretched is called **tension**. As you pull on the rope the particles inside are moving apart. The tension force that you feel is the force of attraction pulling the particles back again.

Elastic bands and bungees

A bungee cord is designed to stretch a very long way. It is made of lots of elastic cords all bound together. If something is **elastic**, it will go back to its original length when you remove the force. The amount that it stretches is called the **extension**.

If something is not elastic, and does not go back to its original length after stretching, we say it is **plastic**.

Stretching a spring

Springs have lots of uses. They are used in spring balances, trampolines, and car suspension systems. It is important to use the right spring for the job, so you need to find out how much a spring will stretch when you apply a force to it.

A bigger force will produce a bigger extension, and if you double the force you double the extension. The extension is **proportional** to the force. This is why we use a spring balance to measure forces. The spring is elastic.

If you keep loading more and more on, eventually the spring will not return to its original length when you remove the force. It has reached its limit, called the **elastic limit**. The spring cannot spring back and it is **permanently extended**.

Elastic materials will break if the force applied to them is too big.

↑ Trampolines need springs that do not stretch very much.

Why do things float?

It seems amazing that something as large and heavy as an oil tanker can float.

When the oil tanker is in the water there is a force pushing up on it called **upthrust**. The upthrust balances the weight so the oil tanker floats. The water particles collide with the bottom of the boat and push it up. You can feel the force of upthrust if you try to push a balloon into a bowl of water.

If you weigh an object underwater it appears to weigh less because of the upthrust. It is much easier to pull yourself out of a swimming pool than it is to pull yourself up when you are not in water.

In this diagram, the upthrust is 4 N. The water level has risen because the weight has displaced some water. The weight of the displaced water is 4 N. This is **Archimedes' principle**. The apparent weight of the object is now 6 N.

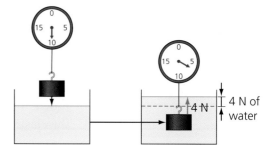

4 N of water

Floating in air

A balloon filled with helium will rise. There is an upwards force on the balloon because there are more collisions between the air and the balloon at the bottom of the balloon than above it. This is because there are more air molecules lower down due to gravity. This upwards force produces an upthrust on the balloon. Its motion is changing so there must be a resultant force on it. When the upthrust is bigger than the weight of the balloon and the air resistance, the balloon will accelerate. When the forces on it are balanced it will move with a steady speed.

Q

1 A spring is 3 cm long. A student hangs a weight of 2 N on it. It is now 4.5 cm long.

   a What is the extension?

   b What would be the extension with a weight of 4 N?

   c What would be the length of the spring with a weight of 6 N?

2 A floating boat weighs 20 000 N. What is the size of the upthrust?

3 Sketch the diagram of the 10 N weight underwater. Draw and label the forces on it.

4 Explain what is meant by the elastic limit of a spring.

5 When people do a bungee jump they are asked how much they weigh. Why?

!

- Springs, ropes, and elastic materials have a tension in them when stretched.

- The extension is proportional to the tension, up to the elastic limit.

- Springs in tension pull back.

- Objects float when the upthrust equals the weight.

Presenting results – tables and graphs

Objectives

- Describe how to present results in tables
- Describe how to draw line graphs
- Explain what is meant by continuous variables

The elastic limit

Suma is on the trampoline. The label says that the weight limit for the trampoline is 1200 N. She works out that anyone with a mass of 120 kg or less can use it. Someone with a mass bigger than 120 kg could be injured on the trampoline. This is because the springs could be permanently extended or even break.

She thinks that for safety, 1200 N must actually be much, much smaller than the elastic limit.

How can I find out the elastic limit of a spring?

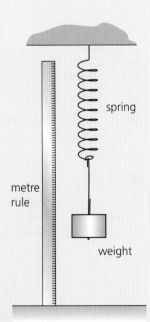

spring

metre rule

weight

Recording the results

Suma collects a spring, some weights, and a metre ruler. She draws a table ready for her results.

Weight (N)	Extension (cm)

Suma measures the length of the spring with no weights on it. This is the original length. There is zero extension, so she records that in her table.

She adds one 1 N weight to the spring and measures the new length. To find the extension she subtracts the original length from the new length. She writes the result in her table.

She continues to add weights and records the results in her table.

Weight (N)	Extension (cm)
0	0
1	1.1
2	2.0
3	2.5
4	4.2
5	5.4
6	6.8
7	8.5
8	11.0

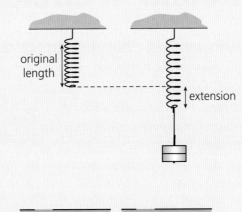

What sort of graph?

Suma knows that to find a pattern in her results, she needs to plot a graph.

Line graphs are a good way of showing information to link two sets of numbers that are both changing continuously. Variables that can take any value like this are called **continuous variables**.

Suma plots the points. She draws a **line of best fit**. This is a smooth line going through or close to as many of the points as possible. The line of best fit shows the pattern or trend in the data. Her line was straight at first, but then curved.

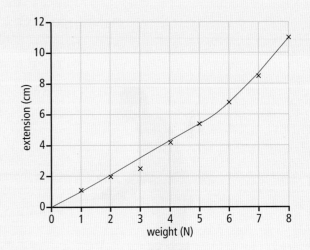

Points that don't fit

One point was far away from the line. This is called an **anomalous point**. Suma probably measured the extension wrongly. All the other results are close to the line.

One good thing about plotting a graph is that you can see if there are any anomalous results and repeat that reading.

Suma noticed that the line started to curve after the weight was 5 N. She took another spring, put 5 N on it and then took it off. The spring returned to its original length. She then tried 6 N. The spring did not go back to its original length. She had found the elastic limit! The elastic limit was 5 N – for weights beyond 5 N this type of spring was permanently extended.

Q

1 What should you write at the top of each column of your results table?

2 Why do you draw a line of best fit rather than joining up all the points with straight lines?

3 Jamila did a different experiment. She put 2 N on lots of different coloured springs and measured the extension of each spring.

   a Draw the table that she would need to record her results.

   b Why could she not draw a line graph of her results?

4 What was the weight for Suma's anomalous result?

5 What mistake might Suma have made to produce the anomalous result?

- A results table shows what you are measuring and the unit.

- If your data are continuous, then you plot a line graph.

- A line of best fit goes through most points and is a smooth line or curve.

Extension 1.10

Round in circles

Objective

- Explain why some objects move in circles

A force that changes direction

Forces can change the shape, speed, or direction of motion of an object. If there is no force on a moving object it will carry on moving in a straight line. If you want to change the direction of motion, you need a force to act.

Ammon is pushing a trolley. The diagram shows how Ammon applies a force to the trolley.

The trolley will move in a straight line. But what will happen if Eboni pushes from the side? The force will change the direction of the motion. The trolley will start to move in a circle.

Ammon pushes → trolley moves this way

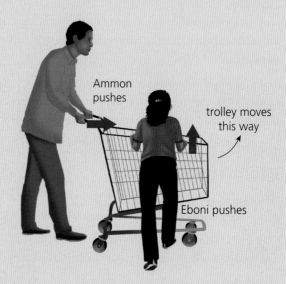

Ammon pushes

trolley moves this way

Eboni pushes

Centripetal force

One of the throwing events in athletics is the hammer throw. The hammer thrower swings the hammer, a big metal ball, around her head on a wire, and then lets it go.

Imagine taking pictures from above as the hammer thrower is swinging the ball. In your first picture, the ball is moving in direction X. In your second picture, it is moving in direction Y. There is a force on the ball from the tension in the wire, which has changed the direction of motion.

If she lets go of the ball then it will move in a direction that is a **tangent** to the circle.

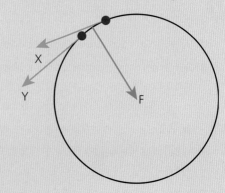

A force that keeps an object moving in a circle is called a **centripetal force**. In the ball on the wire, the centripetal force is provided by the tension in the wire. For cars on the road, the centripetal force is provided by friction between the tyres and the road.

The centripetal force is always directed towards the centre of the circle. This means that the force is always at right angles to the direction of motion of the object.

Centripetal forces in space

The Earth and other planets **orbit** the Sun. The Moon orbits the Earth – it is a natural **satellite**. The Earth also has artificial satellites put into orbit for communication and for studying the surface of the Earth.

The centripetal force that keeps the Moon and satellites in orbit around the Earth is the force of **gravity**. It is gravity that also keeps the planets and asteroids in orbit around the Sun, and keeps moons in orbit around other planets, like Jupiter.

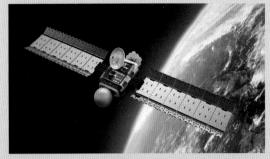

⬆ A satellite needs the force of gravity to stay in orbit around the Earth.

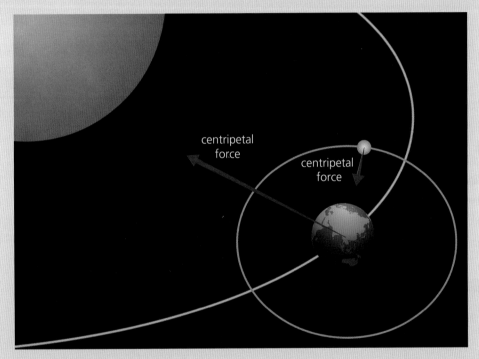

centripetal force

centripetal force

Q

1 Explain what meant by centripetal force.

2 A centripetal force is needed to keep a car moving in part of a circle as it goes around a corner. What provides that force?

3 The Moon orbits the Earth.

  a What is the force that keeps the Moon in orbit around the Earth?

  b Imagine that this force was suddenly switched off. What would happen to the Moon?

4 A student says that there cannot be any force acting on a satellite in orbit if it is moving at a steady speed. What would you say to the student?

- The force that keeps something moving in a circle is the centripetal force.

- The centripetal force can be provided by forces such as gravity, friction, or tension.

- Planets, moons, and satellites stay in orbit because of gravity.

1   What is the difference between **balanced** and **unbalanced** forces?

Copy out the table below and put a tick in the correct column for each statement:

	Forces are balanced	Forces are unbalanced
The object is not moving.		
There is only one force acting on the object.		
The object is accelerating.		
The two forces are the same size but in opposite directions.		
The object is slowing down.		
The two forces are different sizes but in opposite directions.		
The object is moving with a steady speed in a straight line.		

[8]

2   In the pictures below the forces are shown by arrows.

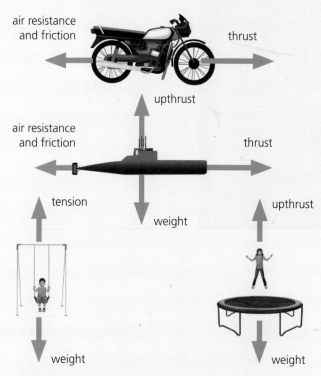

air resistance and friction — thrust

air resistance and friction — thrust
upthrust / weight

tension / weight

upthrust / weight

a   In which pictures are all the forces correctly labelled? [1]

b   In which picture(s) are all the forces balanced? [1]

c   In which picture(s) are all the forces unbalanced? [1]

3   The Tour de France bicycle race takes place every year. It is a road race and in some places the road is flat, but in other places the racers cycle uphill and then downhill.

a   What are the forces acting on the cyclist when he is accelerating downhill? [1]

b   Draw a diagram of a cyclist travelling along a flat road at a steady speed. Add arrows to the diagram and label them. [4]

4   Explain why a streamlined car uses *less* fuel than if it was not streamlined. Use these words in your answer: thrust, air resistance, air molecules, force. [4]

5   Arya is hanging some weights on an elastic band.

a   Here are some tables ready to collect his results. Which table is correct? [1]

Table A:

Weight	Extension

Table B:

Weight (N)	Extension (m²)

Table C:

Mass (kg)	Extension (cm)

Table D:

Weight (N)	Extension (cm)

b   What does the word 'elastic' mean? [1]

c   Arya finds that if he hangs a weight of 2 N on the elastic band the extension is 1.5 cm, and when he hangs a weight of 4 N on it the extension is 4 cm. Would it be a good idea to use an elastic band in a forcemeter instead of a spring. Explain your answer. [2]

6 Tamara measured the length of a spring when various weights were hung on it. Here are her results.

Force on spring (N)	Spring length (cm)
1	6.0
2	8.0
3	10.8
4	12.0
5	14.0

a Draw a graph of spring length (*y*-axis) against force (*x*-axis). Draw a line of best fit through the points.

b How long was the spring with no weights on it?

c What force would be needed to make the spring 8.5 cm long?

d Has the spring been stretched beyond its elastic limit? Explain your answer.

e What would happen if Tamara removed the weights from the spring?

7 Which of these statements about mass and weight are correct? There may be more than one.

A Weight is measured in kilograms.

B Mass is the amount of stuff in an object.

C Weight is a force.

D Mass is measured in newtons. [2]

8 Imagine holding a table tennis ball at the bottom of a beaker of water.

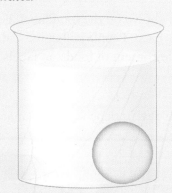

When you let it go it rises to the surface and floats. But if you take it out of the water and let go it falls.

Copy out and complete the statements using these words: bigger than, smaller than, the same as.

a When the ball is accelerating up the weight is _____ the upthrust. [1]

b When the ball is floating the upthrust is _____ the weight. [1]

c When you take the ball out of the beaker and drop it the weight is _____ the air resistance. [1]

9 Alan Shepherd was the fifth person to walk on the Moon. He took a golf club and ball and made a shot. The ball went a lot further than it would on the Earth. Give *two* reasons why. [2]

10 Choose a word from below to match to each of the sentences.

lubrication   drag   elastic limit   proportional

a You put oil on a chain to reduce the force of friction. [1]

b A spring will be permanently extended if you exceed this. [1]

c If you double the force on a spring the extension will double. [1]

d Air resistance and friction are often called this. [1]

11 A person with a mass of 70 kg on Earth has a weight of 700 N. Are each of theses statements true or false?

a If you go to a more massive planet your weight will be bigger

b If you weight on a planet was 300 N that means that the gravitational field is stronger there.

c A planet with a weaker gravitational field would make your weight smaller.

d A planet with a stronger gravitational field would make your mass bigger. [4]

12 Bhaskaracharya and Newton both thought about the Moon.

a What was the question that they both asked about the Moon?

b What was the explanation that they both worked out?

c Why was the discovery of Neptune important?

d Name a scientist who developed their explanations. [4]

What is energy?

You need energy

You need **energy** from food to ride your bike. Energy from **fuels** is needed for transport and to produce electricity. The idea of energy is very useful to help explain what can happen, but it does not explain *why* things happen. For example, fuel allows a car to move, but that doesn't tell you where it is going to go. But energy tells you that if you don't put fuel in the car then it cannot move.

The unit of energy

Energy is not a thing and it is not a substance that moves from one object to another. It is a way of keeping track of a very important quantity, a bit like money.

The unit of energy is the **joule** (**J**). One joule is a very small amount of energy, so we often use **kilojoules** (**kJ**). 1 kJ = 1000 J.

Different foods store different amounts of energy.

Food	Energy (kJ) per 100 g of food
banana	340
beans	400
rice	500
cooked chicken	800
chocolate	1500

The energy stored in food is usually shown on the food label. For a long time a unit called a kilocalorie (kcal) was used. People called it a 'calorie' for short and you still often see the kilocalorie content of foods on labels.

Energy for your body

How much energy do you need each day? It depends on what activities you do.

All activities have an energy cost. Keeping your body warm, breathing, moving, and talking all need energy. Children need energy to grow bigger bones, muscles, and brains.

About three-quarters of the energy that you need every day is for processes in your body like breathing. You then need more energy for all the other activities that you do such as walking, running, or lifting things.

Activity	Energy (kJ) for each minute of activity
sitting	6
standing	7
washing, dressing	15
walking slowly	13
cycling	25
playing football	59
swimming	73

Sportsmen and women need lots of energy. People who take on the challenge of walking to the North or South Pole need even more energy than athletes. This is because as well as walking and carrying their food they need lots of energy to keep warm.

Energy balance

An adult should take in only as much energy as they need for the activities that they do. If they take in more energy than they need, their body stores it as fat for future use. If they eat less than they need, then the body will use energy from its store of fat and they will lose weight.

Energy in fuels

Food is not the only energy store – fuels such as **coal**, **oil**, or wood also provide us with stored energy that we can use.

We can burn wood or coal to heat a room or to cook food. If we use an electric kettle, then the energy needed to boil the water is transferred by electricity. The energy used to generate this electricity may have been stored in a fuel such as coal or oil.

1 Name three fuels.

2 How many joules are there in 200 kJ?

3 Name two things your body needs energy for when you are asleep.

4 Why is it important for young children to take in more energy than they use for their activities each day?

5 For how many minutes would you need to cycle to use up the energy in 100 g of chocolate?

- We use the energy stored in food for all our activities and to stay alive.
- Energy is measured in joules or kilojoules.
- The energy in the food someone eats should equal the energy they need for the activities that they do.

2.2 Energy from the Sun

Where does the energy stored in our food come from?

Objectives

- Explain why the energy in food comes from the Sun
- Describe some methods of generating electricity using the Sun's energy

All of the energy stored in food, like meat, vegetables, or fruit, comes from the Sun. Plants use energy from the Sun to grow, in a process called **photosynthesis**. The plants make roots and leaves. They may also produce flowers, fruit, and seeds. We use some parts of plants for food. You can find out more about photosynthesis in your *Complete Biology for Cambridge Secondary 1* Student book.

The diagram shows how the Sun is the source of energy for rice and for chicken. Rice is one example of a plant that produces grains that we can harvest and eat. Other plants produce other parts that we can eat, including:

- fruit (such as bananas)
- seeds (such as sunflower seeds)
- roots (such as yams).

The energy from the Sun is the source of the energy in the fruit and vegetables that we eat.

The Sun is also the source of energy for meat, because animals get their energy from eating plants. Maize uses energy from the Sun to grow, and animals such as chickens eat the maize. We take in the energy from the animals when we eat chicken or other meat.

Energy in fuels

The Sun is the source of energy in fuels such as wood. The energy that trees use to grow comes from photosynthesis. When we burn wood the energy that we release came from the Sun.

Wood is an example of a **biofuel**. A biofuel is a fuel that we get from living things such as plants or trees. Alcohol is another biofuel that can be made from sugar cane which is grown in countries such as India and Indonesia. Biogas is a biofuel that can be made from the waste from animals and plants.

We use **fossil fuels** for heating, cooking, and transport. Coal is a fossil fuel that formed from trees that were buried under mud millions of years ago. Over time the wood was converted into coal and the mud turned to rock.

↑ The Mangala Area is a major oil field in Rajasthan in India that was not discovered until 1999. It will produce oil for many years.

Oil and **gas** are fossil fuels that formed from sea creatures. They died and were buried under sand or mud at the bottom of the sea. Over millions of years the pressure produced oil and gas from them, and we can now drill down through the rock to find these fuels.

The oil that we find is called **crude oil**, and it is a mixture of many different useful fuels. In a refinery it is separated into individual fuels, such as the petrol (gas or gasoline) that is used in cars.

Wind and water

People have used the energy in the wind for sailing and to turn windmills for thousands of years. We can also use the energy in the wind to generate electricity.

The Sun heats up the land and the oceans by different amounts. This makes the air move, forming the wind. The

↑ Wind turbines produce electricity in India.

wind makes wind turbines turn around. These drive generators, which produce electricity.

The energy from the Sun also evaporates water from the oceans. Water vapour is carried in the air and eventually falls again as rain. We can trap the rainwater behind dams to make artificial lakes. When we release the water we can generate electricity as it falls downhill. This is called **hydroelectricity**.

Using the Sun's energy directly

The energy from the Sun can be used to power **solar cells**. These convert the Sun's energy directly into electricity. They are very useful and can be used on small devices like calculators. Large numbers of solar cells can generate electricity for a house or village.

In **solar panels** water runs through pipes and is heated using the energy from the Sun. This reduces the amount of biofuels or fossil fuels that you need to heat water.

Q

1 Copy and complete this sentence using these words:

   animals   plants   Sun   photosynthesis

   The energy in food comes from the _____. Plants convert the energy in a process called _____. The energy in crude oil and gas comes from the energy in dead _____. The energy in coal comes from dead _____.

2 Explain how the energy in chicken comes from the Sun.

3 What is the difference between a biofuel and a fossil fuel?

4 What is the difference between a solar cell and a solar panel?

5 A student thinks that hydroelectricity means getting energy from water. Is he correct? What would you say to him?

!

- Plants use energy from the Sun in photosynthesis.
- We use energy from the Sun by eating plants for food, and by using fuels such as biofuels and fossil fuels.
- Energy from the Sun can be used to generate electricity using wind, water, or solar cells.

Energy types

Types of energy

Objectives
- Name different types of energy
- Give examples of situations that involve different types of energy

When we move, cook food, or turn a light on, we transfer stored energy from food and fuels. To describe what is happening in processes like these, we can think about energy 'types'. Types of energy give us a short way of describing a store of energy, or a process that transfers energy.

More energy stores

Food and fuels are not the only stores of energy. **Potential energy** is a type of energy that an object has because of its position or shape. A change of position or shape can store energy. Moving objects are also a store of energy.

Food and fuels are chemical stores of energy. We describe this energy as **chemical energy** or **chemical potential energy**. We use the chemical energy from food and in fuels to keep warm.

A mother lifts her baby up from the floor. The baby has gained **gravitational potential energy** (or **GPE** for short). The position of the baby has changed.

If you sit on a bed or a sofa the springs inside it are compressed. The springs are a store of **elastic potential energy** because they have changed shape. Elastic bands and other stretchy materials can also store elastic potential energy.

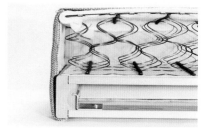

All moving objects have **movement energy**, which is also called **kinetic energy**. A fast-moving car has more movement energy than a slow-moving car.

When fuels burn they heat up the air around them. Energy is transferred to the air. This increases the **thermal energy** in the air. A hot object is a store of thermal energy.

Types of energy transfer

In a torch **electrical energy** is transferred from the battery to the bulb, which produces light. Electric kettles use electrical energy to heat water. Kettles use electrical energy from power stations rather than from batteries.

Energy can be transferred from the Sun or a candle in the form of light waves. We say that the candle or Sun produces **light energy**. We need the light energy to see things.

Musical instruments transfer energy to your ear as sound waves. We say that the guitar produces **sound energy**. Your vocal chords produce sound energy when you talk.

Electrical energy, light energy, and sound energy cannot be stored. We use these names to describe how energy is transferred.

Nuclear energy

The Sun produces the light and thermal energy that we need to live on Earth. Why does the Sun shine? Inside the Sun a process called **nuclear fusion** is happening all the time. Nuclear fusion is a special reaction that produces **nuclear energy**. Hydrogen atoms combine to form helium, and energy is released.

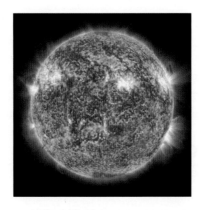

Q

1 Name two sources of light energy.

2 A child picks up a toy from the floor. Which type of energy has the toy gained?

3 A stretched elastic band is a store of which type of energy?

4 A truck moving along the road has kinetic energy. It slows down. Does it have more or less kinetic energy now?

5 **Extension:** a student thinks that the Sun is a huge ball of fire that provides the energy for the Earth. What would you say to them?

- There are different types of energy including chemical, gravitational potential, elastic potential, kinetic, thermal, electrical, light, sound, and nuclear.

35

Energy transfer

Darsh is listening to a battery-powered radio. The radio changes the chemical energy stored in the battery to sound, which travels to his ears.

Energy transfer diagrams

Energy transfer diagrams are a useful way of showing energy transfers. Here is a diagram for a battery-powered radio.

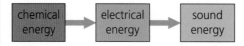

The diagram suggests that the chemical energy is changed into electrical energy which is then changed into sound energy. However, the sound is produced immediately when you switch the radio on. The chemical energy is converted into sound energy straight away. The electrical energy shows the process by which the energy is changed.

Change, transfer, transform?

When you turn on a torch or lift a box, you can think of changes to the types of energy. Sometimes it is helpful to talk about energy being **transferred** from one place to another. You will also see descriptions where the energy is **transformed** from one type to another.

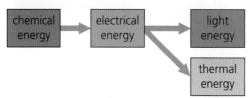

The diagram above is for a light bulb. We can talk about how the energy is transferred as electrical energy, or as light and thermal energy. Alternatively we can talk about how the energy is transformed from chemical energy to light and thermal energy. In either case, the end result is that there is less chemical energy in the battery, and more thermal energy in the surroundings. The light and thermal energy both heat up the surroundings.

Energy changers

There are lots of different devices that change or transfer energy, such as televisions or catapults.

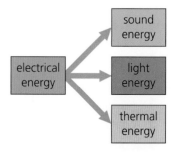

Anika is watching the TV. The TV produces light and sound, and when Anika turns it off she notices that the back of it is warm. The TV has transferred energy to the surroundings in several ways, as shown on the right.

Dewi is playing with a catapult. She puts a ball in the sling, pulls it back, and lets it go. The ball flies through the air. The food Dewi ate gave her the energy she needs to pull back on the catapult. The chemical energy in the food is changed to elastic potential energy in the catapult. When she lets go this is changed to the movement energy of the ball.

Other types of energy change

Here is an energy transfer diagram. Can you guess what the diagram is for?

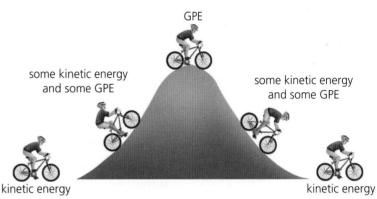

This is an energy transfer diagram for a human being! The amount of energy transferred as thermal energy or kinetic energy depends on what the person is doing. The diagram starts with chemical energy, but we know that the chemical energy in our food comes from the Sun.

Ikenna is cycling along the road. He sees a small hill ahead. He stops pedalling when he gets to the hill. He freewheels up the hill, slowing down as he goes. He gets to the top of the hill and then freewheels down the other side, speeding up as he goes.

There are lots of different energy changes in this story. We can look at different parts of the story. Think about what happens when Ikenna stops pedalling. His kinetic energy is changed to GPE and then back to kinetic energy.

There are lots of examples of this type of energy change, for example a child on a playground swing.

Where does the energy come from?

When you are drawing an energy transfer diagram it is important to think about which part of the process your energy diagram describes. For example, if you are plugging a television into a socket on the wall you might draw an energy transfer diagram as shown on the opposite page. However, the electrical energy has to come from somewhere!

Electricity is generated in a power station, and that might use coal. You will learn more about this in lesson 10.7. Coal is a store of chemical energy. But where did that energy come from? In lesson 2.2 we learned that the energy in coal comes from the Sun. But where does the Sun get its energy from? We learned that there is a process called nuclear fusion in the Sun. So our energy transfer diagrams usually only show a small part of the process.

⬆ Electricity is generated in a power station.

Q

1 What does an energy transfer diagram show?

2 Draw an energy transfer diagram for Dewi using a catapult.

3 Describe the energy transfers in:
   a an electric kettle
   b a torch
   c a child on a swing.

4 A child in a playground climbs up the steps of a slide. She slides down and stops. Draw an energy transfer diagram to show this.

5 **Extension:** draw an energy transfer diagram for the television that starts with the Sun.

● An energy transfer diagram shows the energy changes in a process or device.

Conservation of energy

Objectives

- Know the law of conservation of energy
- Explain how the law applies to different situations

Keeping track of energy

Energy is a bit like money.

Jamal goes out with four coins in his pocket, and uses three of them to buy some food. How many coins are left in his pocket?

You do not need to look in his pocket to know that he has one coin left. If Jamal gets home and no longer has a coin in his pocket then he must have lost it on his way home. He does not think that the coin just disappeared. It must be somewhere. He also knows that he will not find more than one coin because coins cannot just appear.

Energy is like the coins. If you know how much energy you have at the start of a process, then you must have the same amount of energy at the end. Energy cannot disappear and you cannot end up with more energy than you had at the start.

This is called the **law of conservation of energy**. This law says that energy cannot be created or destroyed. It can be transferred in a process. If you know how much energy you have at the start of a process, then you must have this amount of energy at the end. This is one of the most important laws in science.

This law allowed scientists to work out all the different types of energy. If they found a process in which there seemed to be less energy at the end than there was at the beginning, then they knew there must be another type of energy involved.

Useful energy and wasted energy

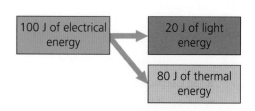

What happens when you turn on a light? The light bulb lights up, and it also heats up. All of the electrical energy is changed into light energy and thermal energy. The **useful energy** is the light energy. We can think of the thermal energy as **wasted energy**. That doesn't mean that it is 'lost', just that it is not the energy that we want. If a light bulb changes 100 J of electrical energy into 20 J of light energy, then 80 J of thermal energy must also be produced.

In all processes some energy is wasted, or changed into a form that we do not want. A car engine changes the chemical energy in the fuel into the movement energy of the car. The engine gets hot, so thermal energy is wasted. When you run, your body changes the chemical energy in food into kinetic energy, but you get hot as well. Thermal energy is wasted, just as in a car.

Thermal energy is produced in many machines because there is friction between moving parts. The friction between moving parts in a bicycle also makes those parts get warm. Thermal energy is transferred to the surroundings.

In some processes the same type of energy is wasted as is useful! In a kettle we want the water to get hot but the air around the kettle heats up as well, this is not useful.

Energy conservation and efficiency

Some light bulbs change most of the electrical energy into light energy. We say that they are **efficient**. This is the energy transfer diagram for an energy-efficient light bulb. Less of the electrical energy is wasted.

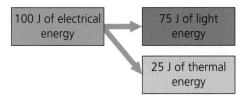

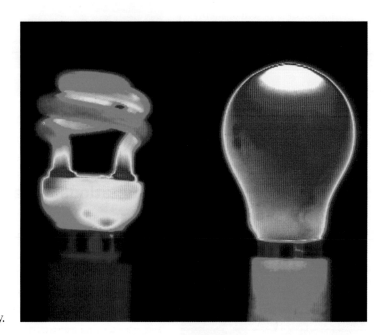

The efficiency of a machine or device tells you how much of the energy that you put in is transferred as useful energy. If a lot of the energy that you put in is wasted, then the machine is not very efficient. Most energy that is wasted is in the form of thermal energy.

1 What is the law of conservation of energy?

2 What is the useful energy and what is the wasted energy in these items?

  a a hairdryer

  b a television

  c a kettle

3 Copy and complete these sentences using the words below.

  wasted   useful   thermal

  A machine is more efficient if it transfers more _____ energy than _____ energy. A lot of the wasted energy in a machine with moving parts is _____ energy due to friction.

4 What would you notice if you stood under an energy-efficient light bulb compared with standing under a normal light bulb?

5 Why do more efficient electrical devices save you money?

- The law of conservation of energy says that energy cannot be created or destroyed.

- In all processes you have the same amount of energy at the end as you had at the beginning.

- In all processes some energy is wasted, usually as thermal energy.

Gravitational potential energy and kinetic energy

Objectives

- Explain what is meant by gravitational potential energy
- Explain what is meant by kinetic energy
- Describe situations that involve energy changes between kinetic energy and gravitational potential energy

This panda is having fun going down a slide at the zoo. As it climbs up the steps the panda gains gravitational potential energy. As it slides down the slide it gains kinetic energy.

Gravitational potential energy

Most cities in the world have very tall buildings. As you travel up in a lift your gravitational potential energy increases. **Gravitational potential energy** or **GPE** is the energy that something has because of its position. Your GPE depends on the distance that you move from the centre of the Earth. You will gain twice as much GPE if you move up two floors than if you move up one floor.

The GPE an object has also depends on its mass. A small boy going up in a lift would gain less GPE than his mother in the same lift. If his mother had twice the mass of the boy she would gain twice as much GPE.

Kinetic energy

Kinetic energy is the energy that something has when it is moving. A lion chasing an antelope has lots of kinetic energy.

- If they are moving at the same speed the lion will have more kinetic energy than the antelope because it has more mass.

- If two lions have the same mass but are running at different speeds the faster lion will have more kinetic energy.

Up and down

Suharto is playing with a ball. He throws the ball upwards. He watches as the ball slows down, stops for an instant, then speeds up again as it falls.

The ball's energy is changing from kinetic energy to GPE and then from GPE back to kinetic energy. The law of conservation of energy tells us that energy cannot be destroyed. The increase in GPE is equal to the decrease in kinetic energy. No energy has been lost if you ignore air resistance.

For the instant that the ball is stationary it has no kinetic energy – it has all been converted into GPE. As the ball falls its GPE is converted back to kinetic energy.

ball stationary

slow speed slowing down | slow speed speeding up

fast speed slowing down | fast speed speeding up

A cricketer scores six runs when he hits the ball over the boundary at the edge of the pitch. When he hits the ball it gains kinetic energy. As the ball moves upwards it gains GPE and loses kinetic energy. As it falls it moves faster, gaining kinetic energy.

Back and forth

A **pendulum** is a ball on a string that swings backwards and forwards. For thousands of years people have designed clocks that use a pendulum. To start the pendulum swinging you pull it back and let it go. When you pull the pendulum back you are also pulling it up, so its GPE increases. When you let it go it falls, and this GPE is converted into kinetic energy. As the ball swings up again its kinetic energy is converted back into GPE.

highest point of swing

no kinetic energy maximum GPE

maximum kinetic energy minimum GPE

A child's father pushes her on a swing. She gains GPE as she swings up. She gains kinetic energy as she swings down. If he stops pushing she will not swing so high each time. This is because other energy changes are happening. Friction is making parts of the swing warm up, and air resistance is making the air heat up. Some kinetic energy is being transferred into the thermal energy of the surroundings.

A rollercoaster

Engineers who design rollercoasters think very carefully about kinetic energy and GPE. They need to know how much energy is needed for passengers to reach the top of the hills. The carriage needs to go up high to gain enough GPE to loop the loop or to go up the next hill.

Q

1 What is the difference between kinetic energy and GPE?

2 A man and his son run to the top of a hill and stop. The mass of the man is bigger than the mass of the boy.

   a While they are both running at the same speed who has more kinetic energy?

   b Would it be possible for them to have the same amount of kinetic energy? How?

   c When they are at the top of the hill who has more GPE?

3 A girl drops a stone down a well and listens for the splash. Draw an energy transfer diagram for this process starting with the girl holding the ball over the well.

4 How is energy 'wasted' on a rollercoaster ride?

- GPE is the energy that something has because of its position.
- Kinetic energy is the energy that something has because of its motion.
- When things move up and down, or swing, energy is changed between GPE and kinetic energy.

2.7 Elastic potential energy

What is elastic potential energy?

Bows and arrows have been used for thousands of years. Scientists in Africa have found arrowheads that date back over 50 000 years.

The archer pulls back on the bowstring and points the arrow at the target. The store of **elastic potential energy** (**EPE**) increases because the shape of the bow has changed.

An elastic band stretches when you pull it. A mattress contains springs that compress when you lie on them. The elastic band and the springs store EPE because their shape has changed.

Objectives

- Explain how the store of elastic potential energy can change
- Describe situations where the store of elastic potential energy increases or decreases

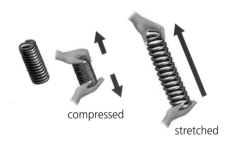

compressed

stretched

Bungee jumping

Grace is doing a bungee jump. She first goes up on a high bridge. She attaches a strong elastic cord to her legs.

When she is standing on the bridge she has GPE. She jumps off and starts to speed up. Her GPE is changed to kinetic energy.

Eventually the bungee cord is straight and starts to stretch. Now some of her kinetic energy changes to EPE stored in the cord. The cord continues to stretch until she stops moving for an instant at the bottom of her fall. At this point all her kinetic energy has been changed into EPE. She then starts to move up again and the EPE is changed back into kinetic energy.

Storing energy in materials

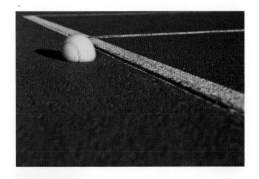

When materials are stretched or compressed their shape changes. We say that they **deform**. When they deform they store EPE. Materials that return to their original shape after being deformed are called **elastic**.

When a ball hits the ground it deforms. Its kinetic energy is changed to EPE when it changes shape. Usually you do not see the ball change shape because it bounces up quickly. If you take a picture when it hits the ground you can see that it deforms.

As it changes back to its original shape and bounces back up, the EPE that has been stored is changed to kinetic energy. Eventually all its kinetic energy is changed to GPE and it reaches the top of its bounce.

If you drop a ball from a height, it will not bounce back up to the same height. This is because not all the kinetic energy that the ball had before the bounce is returned to the ball after the bounce.

When the material deforms its particles change position. This heats the ball up a bit. The thermal energy in the environment increases slightly. The law of conservation of energy says that the total energy before the ball bounces equals the total energy afterwards. If some energy is wasted as thermal energy, then the ball will have less kinetic energy after it bounces than before.

The ball will reach a lower and lower height each time it bounces until it stops. All of the GPE the ball had when you held it up at the start has now been transferred to thermal energy in the environment.

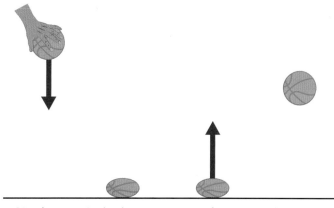

1. GPE changes to kinetic energy.　2. Kinetic energy changes to EPE.　3. EPE changes to kinetic energy.　4. Kinetic energy changes to GPE.

EPE stored in your body

A tendon is a band of tissue that connects a muscle to a bone. One example is the Achilles tendon in your calf. When you walk, this tendon stretches and stores energy and then releases it again.

Some animals, such as kangaroos or frogs, can make huge leaps because of EPE stored in their tendons. They would not be able to leap so high using just their muscles.

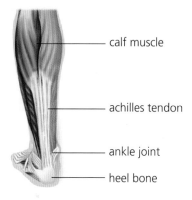

— calf muscle

— achilles tendon

— ankle joint

— heel bone

Q

1 A student is playing with an elastic band, stretching it and letting it go.

  a Describe the energy changes as she stretches the elastic band and lets it go.

  b Explain in terms of energy why the band will go further if she pulls it back more.

2 Look at the pictures of the bungee jumper. She will bounce up and down and eventually come to a stop.

  a Draw an energy transfer diagram for the process from when she jumps off to when she is briefly stationary at the top of her first bounce.

  b Where does all the energy go eventually?

3 Explain why a ball will never bounce higher than the height that you drop it from.

!

- When materials deform they store EPE.
- Some energy is wasted as thermal energy when materials deform.

Suggesting ideas – investigations

Objectives

- Recognise that there are many ways to find answers to questions in science
- Understand how to decide on a question to investigate

Asking questions

Chima is interested in the energy stored in fuels and food. He asks his friends what they would like to know.

How much fuel will Africa need in the future?

Which fuel is the best?

↑ Sanaa

↑ Kojo

Which fuel heats up water the fastest?

Should we use fossil fuels to cook our food?

What types of food do chimpanzees eat?

Which country uses the most coal?

↑ Lumusi

↑ Subira

↑ Thulani

↑ Mosi

Chima looks at all the questions. He wonders how the questions can be answered.

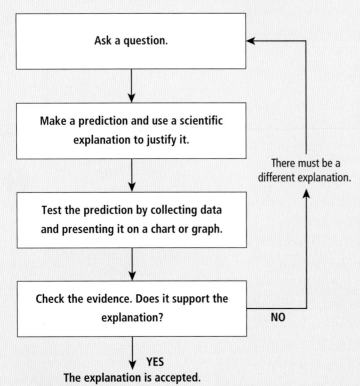

Ask a question.

Make a prediction and use a scientific explanation to justify it.

Test the prediction by collecting data and presenting it on a chart or graph.

Check the evidence. Does it support the explanation?

There must be a different explanation.

NO

YES
The explanation is accepted.

Answering different types of question

There are lots of different ways of finding answers to questions using science. But not every question can be answered using science. This flow chart shows how some scientists think about answering a question.

To answer some questions you need to collect data. There are lots of different ways of collecting data.

A practical investigation

You can collect data by doing an investigation. This is an experiment where you change one variable and keep all the other variables the same. You set up equipment to take measurements to answer the question.

⬆ Lumusi

Lumusi can answer her question by doing an investigation. She can use different fuels to heat up water. She can time how long it takes the water to reach a certain temperature. Then she can use the data to find out which fuel heats up water the fastest.

She will need to decide:

- which fuels to test
- how much water to use and
- the start and end temperatures of the water.

This flow chart shows how she can do her investigation. Scientists may do other types of practical work, for example to find out *if* something happens.

Questions about fuels

Here are some other questions about fuels that you could answer by doing an investigation.

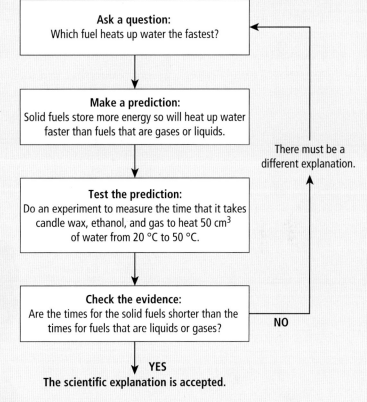

Ask a question:
Which fuel heats up water the fastest?

Make a prediction:
Solid fuels store more energy so will heat up water faster than fuels that are gases or liquids.

Test the prediction:
Do an experiment to measure the time that it takes candle wax, ethanol, and gas to heat 50 cm³ of water from 20 °C to 50 °C.

Check the evidence:
Are the times for the solid fuels shorter than the times for fuels that are liquids or gases?

There must be a different explanation.

NO

YES
The scientific explanation is accepted.

How does the amount of fuel affect the time it takes water to heat up?

Which fuel is the cleanest?

Which type of food stores the most energy?

Q

1 Can you answer the question 'Which fuel is the cheapest?' with a practical investigation? Explain your answer.

2 Here are some data that Lumusi collected in her investigation.

Fuel	Volume of water (cm³)	Start temperature (°C)	End temperature (°C)	Time to reach end temperature (s)
ethanol (liquid)	25	20	30	85
gas	25	20	30	45
candle wax	25	20	30	108

What is the answer to Lumusi's question?

3 Scientists repeat their investigations. Why do they do this?

!

- Scientists ask a question and think creatively about how to collect data.
- Scientists do investigations to collect data to answer a question.
- In an investigation they change one variable at a time and take measurements.

Suggesting ideas – observations and models

Objectives

- Recognise that there are many ways to find the answers to questions in science
- Understand how to decide on a question to investigate

Instead of doing a practical investigation in a laboratory, sometimes scientists use other methods to collect data.

Field study

A **field study** is an investigation into plants or animals in their natural habitat. When doing fieldwork it is important that you make observations without affecting what you are looking at.

Thulani's question could be answered with a field study. To answer this question a scientist would make lots of observations of chimpanzees in their natural habitat. She would collect data and display the data on a chart.

Two girls look at the chart.

What types of food do chimpanzees eat?

↑ Thulani

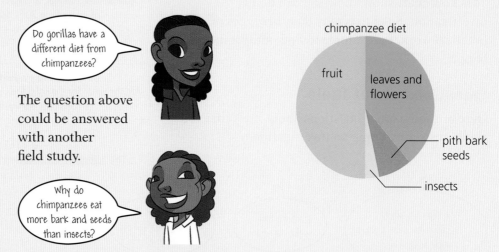

Do gorillas have a different diet from chimpanzees?

The question above could be answered with another field study.

Why do chimpanzees eat more bark and seeds than insects?

chimpanzee diet

fruit

leaves and flowers

pith bark seeds

insects

This question cannot be answered by doing a field study. Scientists could do some experiments to find the benefits of bark, seeds, and insects in a chimpanzee's diet.

Regular observations or measurements

To answer some questions, a scientist might makes lots of observations or measurements. They might do this over a long timescale, or all at the same time but in different locations. Sometimes a scientist collects data that other scientists have already measured before.

Mosi's question could be answered by collecting data from lots of different countries and comparing them.

Some of the most important questions in science are about fuels. We need to find out what will happen to the climate. Scientists are measuring the weather, the sea temperature, and the air temperature to see how they are changing. Many scientists think that there is a link between the climate and the fact that we are burning more fossil fuels now than people did in the past.

Which country uses the most fuel?

↑ Mosi

Making a model

Sometimes it is not possible to do a practical investigation to answer a question – maybe what you are looking at is too big or small or dangerous to experiment on. Scientists have to think creatively about how to find the answer. There are lots of different types of model. Two of them are a physical model and a computer model.

- They can use a physical model. This is useful for very large-scale or small-scale systems. You may have used a physical model of the Earth and the Sun to explain why we have day and night.

- A computer model can help them find the answer. As well as helping to answer the question, a computer model can also be used to predict or to explain.

Sanna's question could be answered by using a computer model. A scientist could use data about the fuels that people in Africa use now. The computer model will help them work out how much fuel they will need in the future.

⬆ Sannaa

How much fuel will Africa need in the future?

Questions that science can't answer

Scientists cannot answer every question. They cannot answer questions when people could disagree, or the answer does not depend on data.

Science cannot answer Subira's question. It could tell you:

- which fuel gives the most energy
- how much carbon dioxide will it produce
- how long will fossil fuel stocks last.

Should we use fossil fuels to cook our food?

⬆ Subira

But people will have an opinion about whether we should use fossil fuels or other fuels. An investigation, a field study, observations, or a model will not tell them whether they are right or wrong. This is because different people will have an opinion about what is really important.

Kojo's question is another one that cannot be answered by science. Again, data can tell you how much energy is stored in fuels, whether they produce harmful gases or smoke, and how safe they are to use. However, you could give two people the same data and they may not agree on the answer.

Which fuel is the best?

⬆ Kojo

1 Two methods of collecting data are practical investigations and field studies.

 a How is a field study different from a practical investigation?

 b How is it the same?

2 Explain why two people may not agree on 'Which is the best fuel?'.

3 Here are some questions. Describe how you might answer them. If you do not think that they can be answered by science or scientists, explain why.

 a Which fuel is the cleanest?

 b How much fuel will we need in the future?

 c Is coal better than gas?

- A field study is a type of investigation.
- Scientists make physical models and computer models to help them answer questions.
- Science or scientists cannot answer every question.

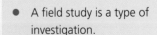

Energy calculations and Sankey diagrams

Efficiency

Melati's mother puts petrol (gas or gasoline) into the car. The petrol is a store of chemical energy. They get in the car and travel home. The car engine gets hot and makes a noise, which is not useful. It also makes the car move, which is useful. Efficiency tells you how much useful energy is transferred compared with the total energy supplied.

Calculating efficiency

You can calculate efficiency using this equation:

$$\text{Efficiency} = \frac{\text{useful energy out (J)}}{\text{total energy supplied (J)}} \times 100\%$$

For every 100 J of stored chemical energy supplied, the car engine transfers 25 J of kinetic energy. We can use the equation to work out the efficiency of the car engine.

$$\text{Efficiency} = \frac{25 \text{ J}}{100 \text{ J}} \times 100\%$$
$$= 25\%$$

This means that 75% or three-quarters of the energy supplied to the car engine is wasted. Thermal energy and sound energy are transferred to the surroundings and heat them up.

Sankey diagrams

A **Sankey diagram** shows the energy transfers taking place in a process. The thickness of the arrow shows the amount of energy transferred. The useful energy is shown by the arrow that points to the right. The wasted energy is shown by arrows that point up or down.

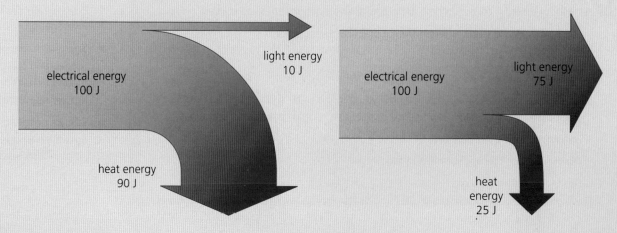

electrical energy
100 J

light energy
10 J

heat energy
90 J

electrical energy
100 J

light energy
75 J

heat energy
25 J

This Sankey diagram shows that for every 100 J of electrical energy supplied to a light bulb, 10 J of useful energy is transferred. The light bulb is not very efficient – most of the energy is wasted. You can use the information on a Sankey diagram to calculate the efficiency.

$$\text{Efficiency} = \frac{10\ \text{J}}{100\ \text{J}} \times 100\%$$
$$= 10\%$$

Fluorescent lights produce a lot of light that spreads out over a wide area. They are far more efficient than a normal light bulb.

You can see that the fluorescent light is more efficient because the light energy arrow is much wider than it is for the normal light bulb.

Sometimes there is more than one type of useful energy, or of wasted energy. A Sankey diagram for a TV would have one arrow for the thermal energy pointing down, but two arrows pointing to the right for light and sound.

Drawing a Sankey diagram

It can be much easier to draw a Sankey diagram if you use squared paper.

You need to decide how much energy each square represents. In this diagram each square represents 20 J. You need to make sure that the total width of the arrows is the same.

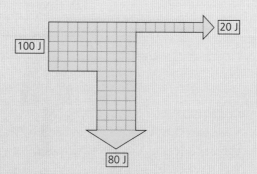

Q

1 A student says that a machine that is not very efficient loses a lot of energy. What would you say to her?

2 Calculate the efficiency of the fluorescent light bulb.

3 Explain why you cannot have a machine that is more than 100% efficient.

4 For every 200 J of energy transferred to a hairdryer, 160 J are transferred to the air to heat it and move it. 35 J are transferred through the case to your hands as thermal energy, and 5 J are transferred to the surroundings as sound.

  a Draw a Sankey diagram for the hairdryer.

  b Calculate the efficiency of the hairdryer.

!

● Efficiency = $\left(\dfrac{\text{useful energy out}}{\text{total energy in}}\right) \times 100\%$

● A Sankey diagram shows how the energy is divided between useful energy and wasted energy in a device or a process.

1 People who do different jobs or sports need different diets.

   a Explain why. [1]

   b List these activities in order. Explain why they are in the order that you have chosen. [1]

     cycling   sitting   walking slowly

2 List these amounts of energy in order from smallest to biggest. [1]

   20 J   20 kJ   2000 kJ   0.2 kJ   2000 J

3 Explain what is meant by:

   a photosynthesis [1]

   b fossil fuel [1]

   c hydroelectricity [1]

   d solar cell. [1]

4 Match each type of energy to its definition. Write out the numbers and the letters.

Definitions:

   1 the energy that something has because of its position

   2 the energy that something has because it has changed shape

   3 energy stored in food or fuels

   4 the energy that something has because it is moving

Words:

   A elastic potential energy

   B kinetic energy

   C GPE

   D chemical energy [4]

5 Which of these statements is *not* correct?

   a Chemical energy is stored in fuels.

   b Thermal energy is what some people call heat.

   c GPE depends on the mass of the object.

   d Kinetic energy does not depend on the mass of the object. [4]

6 A boy is riding his bike. Complete the sentences using the words below. You may need to use each word once, more than once, or not at all.

   thermal   chemical   heat   light   GPE   kinetic

   a The food that he eats for breakfast is a store of _____ energy. [1]

   b The useful energy is _____ energy. [1]

   c The wasted energy is _____. [1]

   d As he moves up a hill and down again the _____ and _____ changes. [1]

   e The _____ energy in the battery stores the energy that he needs for his lights. [1]

7 A car transfers chemical energy in the fuel to kinetic energy so the car moves.

   a Which form or forms of energy are wasted? [1]

   b Draw an energy transfer diagram for the car. [1]

8 Which of these is the correct definition of the law of conservation of energy? [1]

   A You cannot lose energy, but you can get more if you need it.

   B Energy cannot be created but you can lose it.

   C Energy cannot be created or destroyed.

   D Energy can be created and destroyed.

9 Here are four objects.

Which is the useful form of energy for each device? Choose from this list. [4]

thermal   kinetic   sound   light

10 List the following in order of amount of GPE, starting with the one that you think has the least.

A Josie lying in bed

B Josie about to jump out of a plane to do a sky dive

C Josie lying on the floor

D Josie on the top diving board [1]

11 Here are some pictures of a girl jumping on a trampoline. Choose the picture or pictures that matches each statement below.

A just about to hit the mat **B** at the top **C** at the bottom **D** just about to leave the mat

a Here the girl has the most GPE. [1]

b Here the girl has the most kinetic energy. [1]

c Here there is the most EPE stored in the trampoline. [1]

12 A diver stands on a diving board that is 10 m above the surface of a swimming pool. At this height her GPE is 6000 J.

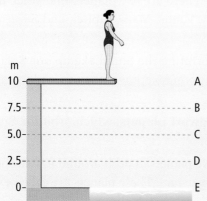

a Copy and complete the table below.

Position	GPE (J)	KE (J)	Total energy (J)
A	6000		
B		1500	
C	3000		
D		4500	
E		6000	

b What law have you used to complete the table?

c What assumption have you made? *Hint:* the diver is moving through the air.

d What happens to the energy when the diver hits the water? [6]

13 Some students are thinking about questions that you can ask. For each question below write down:

a Is it a question that science or scientists can answer? [3]

b If so, which method could you use to collect data to answer it? [3]

These are the questions:

A Which elastic band is strongest?

B Do girls or boys have a better memory?

C How should we grow enough food for everybody in the future?

14 A student has completed an investigation about liquid fuels. This is her prediction:

"I predict that the more fuel I use the faster the water will heat up."

This is a graph of her results:

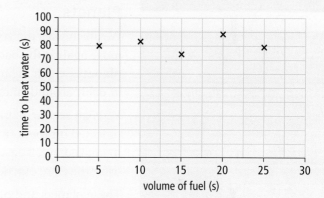

a What has the student missed out in her prediction? [1]

b Do the results support her prediction? Explain your answer. [2]

c Describe a method that the student could have used to get these results. [1]

d Describe the energy changes in this experiment. [1]

e Another student says 'In this experiment the useful energy is thermal energy, and the wasted energy is thermal energy'. Is this correct? Explain your answer. [2]

f Explain why it is difficult to get accurate results when you are using a fuel to heat water. [1]

The night sky

⬆ You can see some objects in the night sky without a telescope. Here you can see the Moon and Venus.

Stars

When you look at the night sky you see millions of dots of light. The dots are **stars**. Stars are huge balls of gas that give out light. When you look at a star it might appear to twinkle – the light seems to flicker. This is not because the star's light is not constant. It is because the light changes direction as it travels through the **atmosphere**.

Stars are the only objects that we see in the night sky that continuously give out light.

Planets

We always see the stars in the same arrangement. Other objects appear to move against the backdrop of the stars. Ancient astronomers named them **planets**, which means 'wandering stars'. Planets are not stars – they are objects in **orbit** around stars. They are made of rock or gas and do not give out light.

You can see the planets and all the other objects in the sky because light is reflected off them into your eyes. There are five planets that you can see with the naked eye. These are Mercury, Venus, Mars, Jupiter, and Saturn. They orbit the Sun, our nearest star.

These planets and other objects orbiting the Sun make up our Solar System. You need a telescope to see Uranus and Neptune, the other planets in our Solar System.

As well as planets there are also **dwarf planets**, which orbit the Sun but are smaller than planets.

Our Sun is not the only star to have planets in orbit around it. Astronomers have found lots of planets that orbit other stars. These planets are called **exoplanets**.

Moons

On a clear night you can see the **Moon**.

The Moon is in orbit around the Earth. It is made of rock. It is the only object beyond the Earth that a person has set foot on.

Most of the other planets in the Solar System have their own **moons**, though you cannot see them without a telescope. The Earth only has one moon, while other planets have

more. Saturn has 60 moons. It is tempting to think that moons are just small planets. However it is not size that determines whether something is a moon or a planet. Titan, one of Saturn's moons, is bigger than the planet Mercury.

A moon is an object that orbits a planet. It is sometimes called a **natural satellite**.

Comets

A **comet** is made of ice and dust like a big, dirty snowball orbiting the Sun. As they get nearer to the Sun it produce a tail that makes a spectacular sight in the sky. Like planets and moons, we see comets because they reflect light from the Sun.

From ancient times people have watched comets in the night sky. They often thought comets were a bad omen. They seem to appear out of nowhere and then disappear. It wasn't until astronomers started to make measurements that they realised that some comets come back again and again. Comet Encke can be seen every three years. Other comets return after hundreds or thousands of years.

Meteors and meteorites

Sometimes you see streaks of light in the night sky. People call them 'shooting stars', but they are not stars. The streak of light is produced when a particle of dust or a piece of rock called a **meteor** enters the Earth's atmosphere and burns up. A piece of rock that survives to reach the ground is called a **meteorite**. You might see lots of meteors together in a meteor shower. This often happens when the Earth moves through the dust left by a comet. Tiny meteorites hit the surface of the Earth all the time. If you go outside about a dozen will land on you every hour.

Man-made objects

There are thousands of man-made or **artificial satellites** in orbit around the Earth. Many are part of the **global positioning system** (**GPS**). This system pinpoints the position of something using signals from lots of satellites. Sometimes you can see these satellites near dawn or dusk when they reflect light from the Sun. The **International Space Station** is the biggest and brightest of all the man-made objects orbiting the Earth.

Q

1 Some of objects in the night sky are in orbit around the Sun, and some are in orbit around planets.

  a Which objects orbit the Sun?

  b Which objects orbit the Earth?

2 Which objects can only be seen because they reflect light from the Sun?

3 What is the difference between a meteor and a meteorite?

4 Why are Uranus and Neptune not called 'naked-eye' planets?

!

- The objects that we can see without a telescope in the night sky include stars, planets, the Moon, comets, meteors, and man-made satellites.

- Stars emit light, but we can see other objects because they reflect light from the Sun.

3.2 Day and night

Objectives

- Explain why the Sun appears to move across the sky.
- Explain why we have day and night.

Sunrise

We are used to the Sun rising every morning and setting every evening. The Sun appears from below the horizon every day at sunrise. It rises in the sky until it reaches the highest point at noon. Then it appears to move down again until sunset.

The Sun is not moving. We are!

Day and night

At any one time, the light from the Sun lights up half of the Earth. The other half of the Earth is in shadow. We can see that from space.

The Earth spins on its **axis**. This means that the area that is in the light changes. **Day** and **night** are caused by the Earth spinning every 24 hours. During the day the part of the Earth that you live on is moving in the light. At night-time it has moved into the dark.

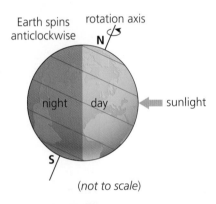

Earth spins anticlockwise — rotation axis — N — night — day — sunlight — S

(*not to scale*)

Looking down from above the north pole of the Earth, the Earth is spinning anticlockwise. This means that the Earth rotates from the west towards the east. So we see the Sun rising in the east and setting in the west. The Sun appears to move across the sky because the Earth is spinning.

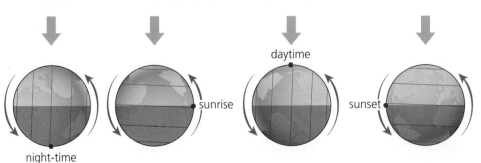

night-time — sunrise — daytime — sunset

Incorrect ideas

It might seem strange to think about the Earth spinning because you cannot feel it moving. You might hear other explanations for day and night.

I think that it is night-time when the Sun goes behind the clouds.

I think that the Sun goes around the Earth. It is night-time when the Sun is on the other side of the Earth.

These ideas are not correct.

Some people think that day and night can be explained by the Earth moving around the Sun. They think the Earth goes around the Sun each day, and that it is night-time when the other side of the Earth is facing the Sun. In this model the Earth is not spinning.

It is true that the Earth moves around the Sun, but it takes one year, not one day!

The spinning Earth

How can we show that the Earth is spinning? There is one very important experiment that shows that the Earth spins. This experiment was first done in Paris, France in 1851. A scientist called Leon Foucault took a very, very long piece of wire, 67 m long. He attached a big ball of metal that had a mass of 28 kg to the wire to make a pendulum. He hung the pendulum from the ceiling of a very tall building so that it was just a metre off the floor. When he started the pendulum swinging it would swing for a very long time because it was so heavy.

People watched the pendulum. Over time the pendulum seemed to change the direction of its swing against a scale on the floor. There was hardly any friction at the point where the pendulum joined the ceiling. The only explanation was that the pendulum was swinging in exactly the same way, but the Earth was spinning beneath it.

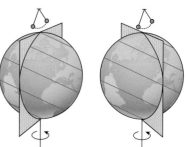

⬆ The pendulum always swings in the same direction, but the scale moves under it as the Earth spins.

1 The Sun rises and sets each day.

   a In which direction does the Sun rise?

   b In which direction does it set?

   c At what time is it highest in the sky?

2 You may remember measuring shadows to see how they change. A student measures the length of a shadow each hour during the day, starting at 8 a.m. and finishing at 8 p.m.

   a When will the shadow be longest? Why?

   b When will the shadow be shortest? Why?

3 A student thinks that the Sun is hidden behind clouds at night. Why is this unlikely?

4 Why did Foucault's pendulum need a very heavy weight on the end?

- We have day and night because the Earth spins on its axis.

- The Earth spins once in 24 hours.

- The Sun appears to move across the sky because the Earth is spinning.

The seasons

Changing day length

Harish lives in Jalandhar, India. It is October and he is watching the Sun setting. It is setting earlier than it did the night before. The days are getting shorter.

Six months later the Sun sets later than it did the night before. The days are getting longer. He wonders why the Sun sets at different times throughout the year.

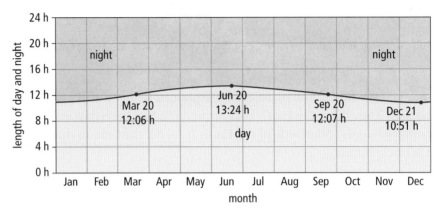

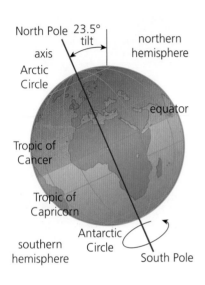

The tilt of the Earth's axis

The Earth orbits the Sun once every **year** or 365 days. We divide the year into **seasons**: spring, summer, autumn, and winter. In some countries near the **poles** there is a big difference in the weather between summer and winter. In summer the days are longer and warmer, and in the winter they are shorter and colder. Other countries near the **equator** do not notice much change.

We can explain why there are seasons by thinking about the tilt of the Earth's axis. The axis is tilted by an angle of 23.5°. This angle of tilt does not change. As the Earth moves around the Sun, for 6 months the North Pole is tilted towards the Sun, and for 6 months it is tilted away from the Sun.

The Earth's tilted axis explains:

● different day lengths in summer and winter

● different temperatures in summer and winter

● the varying height of the Sun in the sky in summer and winter.

spring in
the north and
autumn in the south

Sun

summer in
the north and
winter in the south

winter in
the north and
summer in the south

autumn in
the north and
spring in the south

Day length

Jalandhar is in the **northern hemisphere**. It is summer in Jalandhar when the North Pole is tilted towards the Sun. The Sun stays in the sky for longer each day because the northern hemisphere faces the Sun for longer as the Earth spins. There is a longer day and a shorter night. As Jalandhar goes from summer to winter, sunrise gets later and sunset gets earlier. The days get shorter.

It is winter in the **southern hemisphere** when it is summer in the northern hemisphere. The Sun stays in the sky for a shorter time each day.

The tilt of the axis means that there are regions of the Earth that have very strange summers and winters. In regions in the north inside the Arctic Circle such as Finland, Norway, and Canada, the Sun doesn't set in the summer. These places are known as the 'Land of the Midnight Sun'. The same happens in Antarctica during its summer, but no one lives there to see it. In winter in these places, the Sun does not rise at all. This is called a polar night.

Temperature

The longer days help to explain why it is hotter in the summer. The Sun warms the Earth for longer each day. There is another reason why it is warmer. If the surface of the Earth faces directly towards the Sun, the Sun's rays hit the surface at 90°. If the surface is tilted away at an angle, the Sun's rays are spread out over a bigger area.

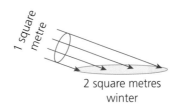
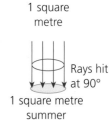

1 square metre

2 square metres
winter

Rays hit at 90°

1 square metre
summer

If the rays hit the surface at 90° then more energy hits the surface per square metre per second. This means that the surface will heat up faster so the air temperature increases more quickly.

Some people think that it is hotter in the summer because the Earth is closer to the Sun in the summer. This is not true. The Earth's orbit is nearly circular. It is slightly closer to the Sun in January each year.

Height of the Sun in the sky

During the day the Sun is highest in the sky at noon. The height the Sun reaches at noon changes over the year. In summer it is higher than it is in winter. This is also because of the Earth's tilted axis.

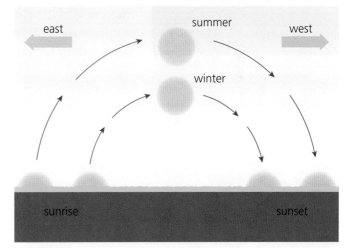

east — summer — west

winter

sunrise

sunset

1 In a country where it is summer, is the Earth is tilted towards the Sun or away from it?

2 Give two reasons why it is hotter in the summer than in the winter.

3 What would you notice about the length of a shadow that a fence post makes at noon in the winter compared with noon in the summer? Explain your answer.

4 The length of a day changes over the year.

   a Explain why.

   b Which month has the shortest day: **i** in the northern **ii** in the southern hemisphere?

5 What would happen to the seasons if the Earth's axis was not tilted?

- The Earth's axis is tilted.
- This explains why it is warmer in summer than in winter, why days are longer in the summer, and why the Sun is higher in the sky in the summer.

Stars

Our Sun

Our **Sun** is a star. You might think that our Sun is the biggest star of all, but it is actually quite small. Some of the bigger stars, such as Arcturus, are hundreds of times bigger than our Sun. Arcturus looks much smaller because it is so much further away.

The stars are shining all the time, but we can only see them when the Sun is low in the sky or at night. This is because the light from the Sun is so bright we cannot see other stars.

The Earth spins

During the night the stars appear to move. If you look at the night sky for a few minutes you cannot see them moving. If you take a photograph over several hours the stars make tracks. From the northern hemisphere and looking north, the tracks form circles.

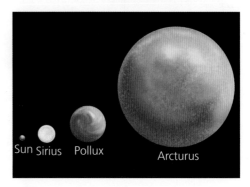

Sun Sirius Pollux Arcturus

The stars are not really moving in circles. They appear to be because of the motion of the Earth, not the motion of the stars. It happens because the Earth is spinning on its axis.

If you take the photograph from the northern hemisphere, the star closest to the centre of the circles is the northern pole star, or Polaris. It is called a pole star because it is directly overhead when you are standing at the North Pole. It is lined up with the axis of rotation of the Earth.

You get the same effect if you take a photograph in the southern hemisphere looking south. This time the star closest to the centre is the southern pole star, Sigma Octantis. All the other stars apart from the north and south pole stars appear to move during the night. Sailors have used the pole stars for navigation because they do not appear to move.

Constellations

People have been trying to make sense of all the stars for thousands of years. A collection of stars that makes a pattern is called a **constellation**, such as the Plough (Ursa Major). It is easier to describe stars in constellations rather than individual stars.

If you live in the northern hemisphere you will see different stars from someone who lives in the southern hemisphere. Cassiopeia is a constellation that you can see from the northern hemisphere, while you can see Crux from the southern hemisphere.

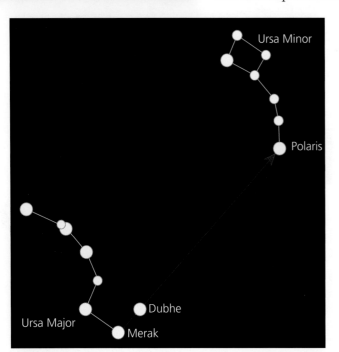

Ursa Minor

Polaris

Dubhe

Ursa Major

Merak

↑ You can use constellations to find particular stars in the sky, such as the northern pole star.

The Earth orbits the Sun

You see different stars in the night sky in December from those in June. The half of the Earth that is dark is facing the opposite way in the sky in June and December. This is because the Earth is in orbit around the Sun.

You can only see the stars in the direction facing away from the Sun. The constellations that you saw six months ago are blocked by the Sun.

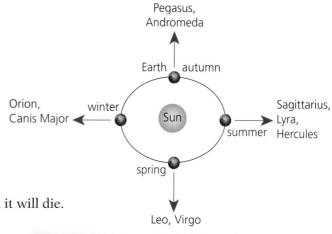

The life and death of a star

Our Sun has a **life cycle**, as an animal does. It is born and it will die.

Like all stars the Sun was born from a cloud of gas and dust called a **nebula**. Gravity pulled the gas together. **Hydrogen** gas is used in nuclear fusion reactions to produce the energy that makes the Sun shine. This is the stage where our Sun is now. It is a **main sequence** star. It will carry on shining like this for another 5 billion years.

Eventually all of the hydrogen gas will be used up and it will go through the final stages of its life. It will become a **red giant** and will swallow up Mercury, Venus, and possibly Earth as well. It will become far too hot for humans to survive on Earth. At this point the outer layers of the Sun will be thrown out into space to form clouds of gas. After that it will shrink back and become a hot **white dwarf**, and then cool down to become an invisible **black dwarf** star.

A star that is much bigger than our Sun is called a **massive** star. It will turn into a **red supergiant**, explode to form a **supernova**, and then form a **neutron star** or a **black hole**.

⬆ Crux – the Southern Cross.

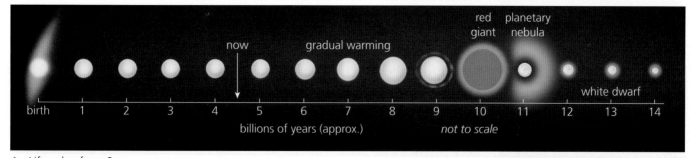

⬆ Lifecycle of our Sun.

Q

1 Why can you not see stars during the day?

2 What is a constellation?

3 Why do we see different stars in the summer and the winter?

4 Why could sailors not use stars in constellations to navigate?

5 **Extension:** roughly how long is the entire life cycle of a star like our Sun? Explain how you worked it out.

!

- The Earth spins on its axis, so the stars appear to move during the night.

- The Earth orbits the Sun, so we see different stars in the summer and the winter.

- You can see different constellations in the northern and southern hemispheres.

Our Solar System

Our planet is Earth. It is one of eight planets in the **Solar System**. A solar system is a collection of planets in orbit around a star. The star at the centre of our Solar System is the Sun.

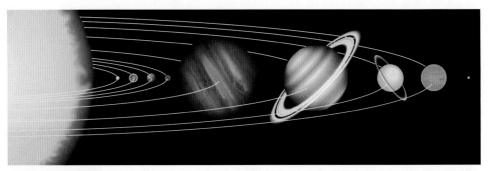

(*not to scale*)

Our Solar System was formed about 5 billion years ago from dust and gas. Gravity pulled the dust and gas together to make the Sun and all the planets. There are four **inner planets**, an **asteroid belt**, and four **outer planets**. There are also dwarf planets in the asteroid belt and beyond the outer planets.

The inner planets

The inner planets are **Mercury**, **Venus**, Earth, and **Mars**. They are often called terrestrial planets because they are made of rock.

Mercury is closest to the Sun. Temperatures on Mercury can reach 430 °C. At night it cools to –170 °C.

Venus reflects a lot of light from the Sun. In some positions it can reflect so much light that it looks like a star. You might hear it called the Morning Star or the Evening Star, but it is not a star. We see it because it reflects light from the Sun. Unlike Mercury, the planet Venus has an atmosphere. It is mainly carbon dioxide, which is a **greenhouse gas**. The atmosphere traps energy from the Sun, so the surface of Venus is hotter than the surface of Mercury even though it is further from the Sun. The average surface temperature of Venus is over 460 °C.

We think that Mars is the only other planet in the Solar System that may once have had life. The *Curiosity* mission was launched in 2012 to see if Mars could support life. There are features on the surface of Mars that suggest that water once flowed there. Mars is known as the Red Planet because the surface is made of a red mineral that contains iron. You will learn more about conditions for life in your *Complete Biology for Cambridge Secondary 1* Student book.

The asteroid belt

Between Mars and Jupiter is the asteroid belt. **Asteroids** are pieces of planets or moons, or lumps of rock left over from when the Solar System was formed. There are millions of asteroids in the asteroid belt, and they orbit the Sun like the planets do. The largest asteroid, Ceres, is also a dwarf planet. It is one-third the size of the Moon.

The outer planets

The four outer planets, **Jupiter**, **Saturn**, **Uranus**, and **Neptune**, are all much bigger than the inner planets.

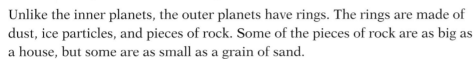

Jupiter and Saturn are called gas giants because they are mainly made of hydrogen and helium gas.

Uranus and Neptune are called ice giants. They are made of ice and rock with an outer layer of gas. They have the coldest atmospheres in the Solar System.

Unlike the inner planets, the outer planets have rings. The rings are made of dust, ice particles, and pieces of rock. Some of the pieces of rock are as big as a house, but some are as small as a grain of sand.

Dwarf planets

Until 2006 **Pluto** was a planet. Then **astronomers** found other objects like Pluto beyond Neptune, so they started calling Pluto a dwarf planet. There are five dwarf planets in the Solar System including Ceres, which is an asteroid.

Size and distance

Name	Distance from the Sun (million km)	Diameter (km)
Sun		1 391 000
Mercury	58	4879
Venus	108	12 104
Earth	150	12 756
Mars	228	6787
Jupiter	778	142 800
Saturn	1427	120 660
Uranus	2871	51 118
Neptune	4498	49 528

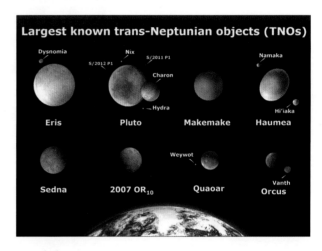

Largest known trans-Neptunian objects (TNOs)

Sometimes it is difficult to imagine the scale of the Solar System and the sizes of the planets. We can use analogies to help us to understand.

If the Sun was my size, then Jupiter would be the size of my head and the Earth would be less than the size of my smallest fingernail.

I am I metre away from you. If you were the Sun and I was the Earth, then Neptune would be 30 metres away from you.

Q

1 Why do scientists think that there could once have been life on Mars?

2 Look at the table of distances from the Sun and diameters.

   a List the planets in order of size.

   b Roughly how much bigger is the Sun than the largest planet?

3 How do the inner planets compare with the outer planets in terms of:

   a their size

   b what they are made of?

4 A student thinks that the outer planets have longer years than the inner planets. Do you agree? Explain your answer.

- The Solar System contains four inner and four outer planets.

- Asteroids orbit the Sun between Mars and Jupiter.

The Moon

Moonlight

Unlike the Sun you can see the Moon in the sky during the day *and* during the night. You see the Moon because light from the Sun is reflected into your eyes. Moonlight is actually reflected light, not light that is produced by the Moon. The Moon does not give out light.

The phases of the Moon

The Moon orbits the Earth every 27.3 days. If you take a picture of the Moon each night you would see that its shape changes in a regular way. These are called the **phases of the Moon**.

The light from the Sun lights up half of the Moon, just as it lights up half of the Earth or any other planet or moon in the Solar System. The Moon appears to change shape because it is lit by the Sun and it is moving around the Earth.

- When the Moon is between the Earth and the Sun, the Moon is in shadow. It appears dark or black and we call this a **new moon**.

- In position X we see half of the Moon lit up. This is the **first quarter**.

- When the Moon is on the opposite side of the Earth to the Sun, we see the whole face of the Moon lit up. We call this a **full moon**.

- In position Y we again see half of the Moon lit up. This is the **last quarter**.

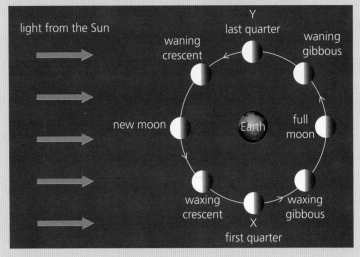

↑ Why we see the phases of the Moon. The Earth and the Moon are not to scale.

As the Moon moves between a new moon and a full moon we say that it is **waxing**. We see more and more of the Moon lit up. As it moves from a full moon to a new moon we say it is **waning**. Less and less of the Moon appears to be lit up.

Between a new moon and the first or last quarter we only see a small part of the Moon lit up. This is a **crescent moon**. Between the first or last quarter and the full moon most of the Moon is lit up. This is a **gibbous moon**.

Eclipses

When an object moves in front of a source of light, a shadow is formed. This can happen when the Moon or the Earth moves in front of the Sun.

When the Moon blocks the light from the Sun, this can cause a **solar eclipse** on Earth. It is called a *solar* eclipse because the light from the *Sun* is blocked.

↑ The Moon casts a shadow on the Earth to produce a solar eclipse.

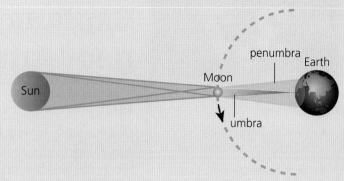

There is a region of deep shadow where the Moon blocks the light from the Sun and this is where you see a **total eclipse**. Outside that region you see a **partial eclipse**. The deep shadow is called the **umbra** and the lighter shadow is called the **penumbra**.

We see eclipses because the Moon happens to be the right size and distance away from the Earth to exactly cover up the Sun. This is a complete accident. You can find out more about shadows on page 110.

When the Earth is between the Sun and the Moon we have a **lunar eclipse**. It is called a *lunar* eclipse because we do not see any of the light that is reflected from the *Moon*. The Moon is in the shadow of the Earth.

Why are eclipses rare?

You might expect to see a solar eclipse every 27.3 days because the Moon orbits the Earth in that time. In fact total eclipses do not happen very often. This is because the orbit of the Moon is tilted slightly. The Earth, Sun, and Moon need to be lined up for an eclipse to happen, so eclipses are rare.

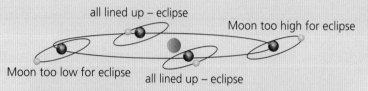

If the orbit of the Moon was not tilted then the Earth would block out the light and we would not see a full moon.

The far side of the Moon

We only ever see one side of the Moon. The other side is sometimes called the **dark side of the Moon** because we cannot see it from Earth, not because it is dark. Half the Moon is lit up all the time. The first photograph of the far side of the Moon was taken by a space probe in 1959. It was not until the *Apollo 8* space mission to the Moon in 1968 that a human being saw the surface.

We do not see the other side of the Moon because the Moon spins on its axis as it orbits the Earth. It happens to take exactly the same time to spin once as it does to orbit the Earth. This means that we always see the same side facing towards the Earth.

Q

1 Write down the names of all of the phases of the Moon in order, starting with a new moon.

2 Draw a diagram of the Sun, Earth, and Moon showing how a lunar eclipse happens.

3 Describe one difference that we would notice if the orbit of the Moon was not tilted?

4 We only ever see one side of the Moon.

 a Why is this?

 b How is it possible to take a photograph of the 'dark side of the Moon' if it is dark?

!

- The Moon has phases such as a full moon, half moon, and new moon.

- We see phases because the Moon orbits the Earth.

- We only see one side of the Moon because the Moon spins once every 28 days.

Explanations – geocentric model

Objectives

- Describe how scientific explanations are developed
- Understand the importance of evidence

Past times

Imagine that you were living 2000 years ago. Many things about your everyday life would be different, but some things would be the same. When you looked up you would see the stars and planets change position just as they do today.

People have always asked **questions** about the way that the Sun, stars, and planets moved. Why do they move that way?

Astronomical evidence

In the ancient world people told stories to their children. The stories explained why things happened, like the sun rising and setting, and seasons of the year. People remembered the stories and told them to their children.

There is a difference between stories and scientific **explanations**. Scientists ask **questions**, for example, 'Why do the stars and planets move in the way that they do?'. Then they find **evidence** by making observations or measurements. They develop explanations to account for the evidence. Sometimes the explanations use a **model**. A model is a physical or mathematical way of explaining how something works, like a model of the Solar System.

The invention of numbers and of writing were important for developing explanations. Without numbers people could not make measurements. If they could not write down their measurements, they could not look for patterns.

The geocentric model

Ancient Greek astronomers believed that the Earth was at the centre of the **Universe**. The Universe is everything that there is. The model that they developed with the Earth at the centre is called the **geocentric model**. 'Geo' means Earth in the Greek language.

Plato and Aristotle were Greek astronomers who lived about 2400 years ago. They developed the geocentric model. In this model the Earth was not moving, but the stars and planets moved around the Earth. The Greek astronomers thought that there were seven planets: Mercury, Venus, Mars, Jupiter, Saturn, the Sun, and the Moon, and then there were stars beyond them.

In the geocentric model all these objects moved on spheres around the Earth. The Moon was on the sphere that was closest to the Earth. The stars were on a sphere furthest from the Earth. All the spheres moved at a steady speed, and all the objects, including the Earth, were spherical.

Light could travel through the spheres, called 'crystalline spheres'.

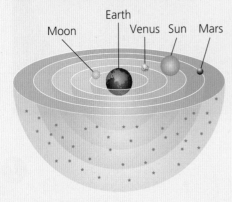

Moon Earth Venus Sun Mars

Retrograde motion

The model explained a lot of the observations, but it did not explain one important observation. The planets seemed to wander across the sky. They did not always go in the same direction.

Ptolemy was another Greek astronomer who lived about 1800 years ago. He developed the model to explain the motion of the planets. He added little circles on top of the bigger circles that were the paths of the planets. He worked out how big they needed to be to explain the 'backward' motion of the planets.

Ptolemy's model explained the motion of the stars and the planets. It also predicted their future positions. He wrote about the model in a book that was used all over the world for hundreds of years.

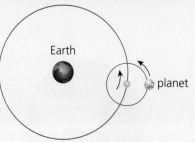

Evidence for the geocentric model

One of the most obvious observations about the Earth is that the ground that you are standing on does not feel as if it is moving. This supported the ideas that the Earth was at the centre of the Solar System.

Aristotle thought that the Earth was at the centre of the Universe because if you drop something it falls towards the Earth. He believed that all objects fell towards the centre of the Universe, so the Earth had to be at the centre.

Ptolemy thought that the Earth was the centre of the Universe because half the stars were above the horizon and half were below the horizon at any time. If the Earth was not at the centre the numbers of stars above and below the horizon would not be the same.

If the Earth was moving then a star would appear to move between January and July as the Earth moves around the Sun. The stars *do* move but they are so far away you cannot see that movement without a telescope.

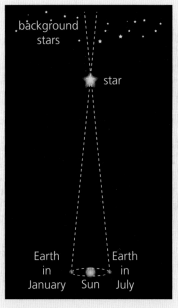

View from Earth:

More evidence changes explanations

Ptolemy's geocentric model was very successful because it could explain lots of observations and make predictions. It was a very popular model because it fitted in with what people could see. It would take evidence from things that they *couldn't* see to change the scientific explanation.

Q

1 Why was the invention of writing important for the development of scientific explanations?

2 Why did people find it easy to believe the geocentric model?

3 How was the geocentric model changed to explain the observations that the planets seemed to wander across the sky?

!

- Scientists collect data and look for patterns in the data.
- They develop explanations for those patterns.
- They look for further evidence for support their explanation.

Explanations – heliocentric model

Challenging the geocentric model

People who lived before 1600 had only ever seen five planets. They saw those planets and the Sun move across the sky. They were happy with the explanation that astronomical objects were on celestial spheres. To change people's view of the Solar System needed new evidence. That new evidence came with the invention of the **telescope**.

Copernicus

Nicholas Copernicus was born in 1473 in Poland, so he lived before telescopes were invented. He is famous for developing the **heliocentric model** of the Solar System, but he did not invent the idea himself. 'Helio' means 'Sun' in the Greek language. He used ideas from other astronomers, such as the Greek astronomer Aristarchus who had produced a model with the Sun at the centre about 1200 years earlier. Islamic astronomers, such as Al-Biruni, had also suggested heliocentric models.

Copernicus promoted the heliocentric model in Europe. In the heliocentric model the Sun is at the centre of the Solar System and all the planets are in orbits around the Sun. This model was able to explain the motion of the planets much more simply. In this model the Moon was in orbit around the Earth.

The fact that the heliocentric model was much simpler than the geocentric model is very important. Scientists always look for the simplest explanation of their observations. If explanations begin to get very complicated, then they will look for a simpler explanation.

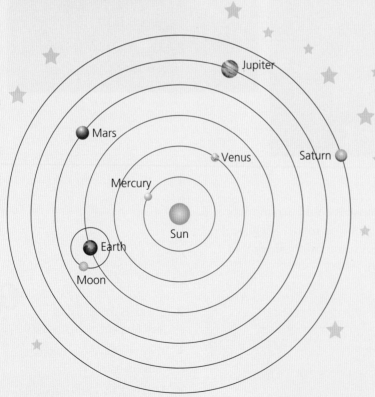

It was very difficult for Copernicus to openly discuss his heliocentric model. Everyone believed the Earth to be at the centre of the Universe at the time. This was an important part of religious beliefs. Copernicus was worried that he would be persecuted if he published his ideas. His book about the heliocentric model was published as he died. It was published in Latin, which was not a language that many people could understand.

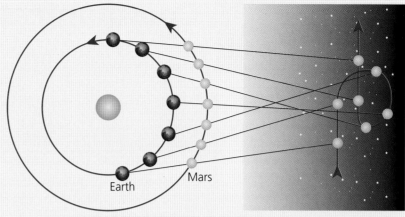

Earth   Mars

⬆ The planets show retrograde motion when viewed from Earth because they are orbiting the Sun at a different speed.

Galileo

Galileo Galilei was an Italian mathematician and scientist. The telescope had been invented in 1600 in the Netherlands, and Galileo made his own version in 1609. He took his telescope out one night and pointed it at the planet Jupiter. The observations that he made are some of the most important of all astronomical discoveries. He saw four objects that seemed to be in orbit around Jupiter, not around the Earth. These were the four biggest moons of Jupiter: Ganymede, Callisto, Europa, and Io. These moons are now called the Galilean satellites in his honour.

This was evidence that not all the objects in the Universe orbited the Earth. It supported the heliocentric model. Galileo thought that the moons orbited Jupiter just like the planets orbited the Sun.

Copernicus did not discuss the heliocentric model in public, but Galileo did. Galileo also used his telescope to make observations of the Sun. He observed dark spots on the surface of the Sun called **sunspots**. The observations showed that astronomical objects such as the Sun were not perfect, as most people believed. His book *The Starry Messenger* is about the moons of Jupiter, our Moon, and his observations of stars. His work on sunspots was published later in *Letters on Sunspots*.

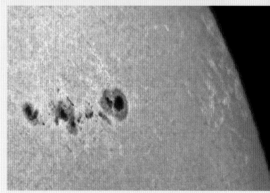

Church leaders in Rome were not happy with Galileo. They tried to make him take back the claim that the Sun was the centre of the Solar System. He refused, and published his ideas in 1632. His book showed how the heliocentric model explains the evidence more simply than the geocentric model. The book was published in Italian, which was easily understood by a lot of people. Again the Church ordered him to take back his claim, and he was kept imprisoned in his house until he died in 1642. After his death, the heliocentric model gradually became accepted.

Q

1 What is the difference between a geocentric model and a heliocentric model?

2 Which piece of evidence was important for overturning the geocentric model?

3 What observation was simpler to explain with the heliocentric model than with the geocentric model?

- New evidence is needed to change explanations.

- Sometimes it is difficult for people to accept new explanations.

Communicating ideas

Scientists in countries all over the world asked question about stars and planets, collected evidence, and developed explanations. Sometimes similar explanations were developed in different places or at different times.

Objective

- Describe how ideas about the motion of the stars and the planets developed in India, Africa, and Islamic countries

There were several reasons why scientists might not have known that an explanation had already been developed by someone else. It may not have been published in a book, or it might have been written in a book that hadn't reached them yet, or it might have been written in another language. Scientists were not able to **communicate** with each other as they do today.

Some scientists did know about the work of others and tried to improve on their models and explanations.

Astronomy in India

The Indian astronomer Aryabhata was born in Bihar, India in the year 476, about 1000 years before Copernicus. He thought that the motion of the stars was because the Earth was spinning on its axis, not because the stars were moving. The Earth spinning also explained why we have day and night.

Another Indian mathematician–astronomer Brahmagupta described a law of gravity about 1000 years before Newton. He worked out a method for calculating the positions of planets, and when they rise and set.

All of these ideas explained the evidence from the observations that astronomers made.

At the time some people thought that eclipses happened because dragons or demons attack the Sun during an eclipse, or that the Sun was swallowed by the gods. Aryabhata showed that his model could predict when lunar and solar eclipses would happen.

Astronomy in Africa

Thousands of years ago the Egyptians used instruments such as water clocks for measuring time. They used merkhets for measuring the position of stars and planets in the sky.

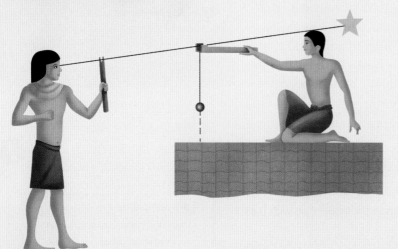

They made a huge number of observations and worked out a calendar with 365 days. This was not the same as saying that the Earth goes around the Sun in 365 days. It was an explanation of the observation that the seasons change over a year.

One of the oldest astronomical sites is in southern Egypt at Nabta Playa. Scientists think that people put stones in places that would line up with where certain stars would rise.

People were probably making measurements of the stars many thousands of years ago in West Africa. Another ancient astronomical device is at Namoratunga, on the west side of Lake Turkana in Kenya. Scientists think that people put the stones there in 300 BC and tilted them to line up with stars at particular times in their calendar.

Islamic astronomy

The earliest records of astronomical observations came from Babylonia, a region that is now in Iraq. People there recorded the positions of planets, and when eclipses happened. Their observations were used by Greek and Indian astronomers.

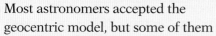

Over 1000 years ago Islamic astronomers translated the books of Greek astronomers. They wanted to find out about their models to explain the motion of the planets and stars. They tried to improve the models to make them fit the observations better, and so they could make predictions.

Most astronomers accepted the geocentric model, but some of them asked questions about it. One of those astronomers was Al-Haytham, who was born in Basra in Iraq. In about the year 1000 he wrote about Ptolemy's geocentric model of the Universe. Later he thought that the Earth was spinning on its axis. At about the same time Al-Biruni, a scientist born in Uzbekistan, thought that it was not possible to prove that the Earth was moving. He developed models such as the one on the right to explain the phases of the Moon.

One of the big achievements of Islamic science was that they built lots of observatories and made new instruments, such as the astrolabe. With these instruments they could make very precise measurements of the stars and planets. They made very detailed charts of the positions of the stars.

Q

1 Could astronomers have shown that the Earth was spinning? (*Hint*: look at section 3.2 on day and night.)

2 Make a list of the astronomical instruments that astronomers developed.

3 Why was it important to develop new astronomical instruments?

4 When a big new observatory was built in a city, lots of astronomers would come to work there. Why would this help them to develop explanations?

!

- Astronomers in India, Africa, and Islamic countries developed their own explanations, or improved the explanations of others.

- Astronomers developed new instruments to help them make better observations.

Beyond our Solar System

Objective

■ Describe what is outside our Solar System

Galaxies

If you look at the night sky you might see a faint white band of stars. This is stars in our **galaxy**, the **Milky Way**. A galaxy is a collection of stars, and there are thousands of millions of stars in our galaxy.

Our galaxy is just one of thousands of millions of galaxies in the Universe. **Andromeda** is another of those galaxies.

⬆ One of our neighbouring galaxies is the Andromeda galaxy.

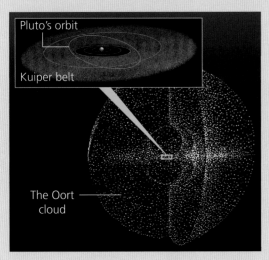

Pluto's orbit

Kuiper belt

The Oort cloud

Outside the Solar System

The *Voyager 1* space probe was launched from Earth in 1977. For over 30 years it has travelled away from the Earth, taking photographs of the planets on its journey. It was designed to study Jupiter and Saturn, but is now the furthest man-made object that there is. On the edge of the Solar System is the **Kuiper belt**, which extends from Pluto's orbit outwards. Beyond the Kuiper belt is the **Oort cloud**. Scientists think that comets, asteroids, and dwarf planets might come from these regions. As Voyager leaves the Solar System it will enter **interstellar space**.

Our nearest star

The nearest star to our Sun is **Proxima Centauri**. It was discovered in 1915 and is only visible with a telescope. You can only usually see it from the southern hemisphere. It has a diameter seven times smaller than our Sun's diameter.

The Oort cloud extends to about one-quarter the distance from our Sun to Proxima Centauri. From the edge of the Oort cloud to Proxima Centauri is gas and dust left over from the formation of our Solar System and other stars.

The Milky Way

Our Sun and its Solar System are part of the Milky Way galaxy. Most of the stars that you can see are stars in the Milky Way. Astronomers think our Sun and the other stars in our galaxy are all orbiting a huge **black hole** at the centre.

The Milky Way is a spiral galaxy. It is shaped like a disc with a bulge in the middle. Our Solar System is about two-thirds of the way out from the centre. The images that we have of the Milky Way are models – they are not pictures that anyone could take using a camera. Scientists have used observations with telescopes to work out the shape of our galaxy.

The hazy white band that we see in the sky is the concentration of stars that we see when we look along the disc. When we look in other directions there aren't so many stars.

Other galaxies

Not everything that you see in the night sky is a star in the Milky Way. Some of the bright, fuzzy objects are other galaxies. One of the most spectacular galaxies is the Andromeda galaxy, which is the nearest spiral galaxy to our own. You can also see two of our other neighbouring galaxies, the Large and Small Magellanic Clouds. You need a telescope to see the galaxies in any detail. With the naked eye they look like fuzzy discs.

It wasn't until 1923 that astronomers could work out that the bright fuzzy spots were other galaxies rather than distant stars within our own galaxy. The word 'galaxy' comes from the Greek word for 'milky'.

Astronomers have used telescopes to identify billions of galaxies but it is not possible to count all of them. They have seen that there are lots of different shapes that galaxies can have, not just spiral.

↑ The Sombrero galaxy.

The Universe

The **Universe** is everything that exists. All the stars, planets, moons, comets, asteroids, meteors, and everything else make up the Universe, including you! Some people ask what is outside the Universe. This is not a question that science can answer because there is no 'outside'. The Universe is **infinite**, which means that there is no limit or edge to it.

We have used a very powerful space telescope, the **Hubble Space Telescope**, to look at galaxies that are more than 12 billion light years away. This is like looking back in time.

How do we know?

Scientists say that there are more stars than there are grains of sand on all the beaches in the world. It is impossible to count all the grains of sand, so how do they know?

They can count the grains of sand in a much smaller volume, say one cubic centimetre. Then they can estimate how many cubic centimetres there might be on all the beaches, and multiply the two quantities together.

This is similar to the way astronomers count stars. They look at a small section of the sky and count the stars in it. They multiply that number by the number of sections to find the total number of stars.

We can never know the exact number of stars or of galaxies. Scientists make estimates of those numbers, just like the number of grains of sand, or even the number of people or animals on the planet.

Q

1. What is the difference between a solar system and a galaxy?

2. Imagine you are in a spacecraft moving away from the Earth. You reach the orbit of Neptune. Copy and complete these sentences.

   As I leave the Solar System I enter the _____ _____. Beyond that I am in the _____ _____ where we think some comets come from. If I carried on I would travel out of our _____ , which is called the Milky Way.

3. Why is it not possible to count all the stars in the Milky Way?

!
- A galaxy is a collection of stars and planets.
- Our galaxy is the Milky Way.
- The Universe is everything that there is.

71

Using secondary sources

Eko and Yuda are looking at pictures of the planets.

> I wonder if there is a link between the length of a day and the length of a year on each planet.

> I wonder if there are seasons on other planets.

Objectives

- Know the difference between primary and secondary sources of data
- Name some secondary sources
- Use information from secondary sources to answer questions

They cannot answer these questions by doing experiments in a laboratory.

Primary and secondary sources

In an experiment you take measurements or make observations. These observations or measurements are called **data**. Scientists record their data in tables and display them using charts or graphs. An experiment is an example of a primary source of data in science. There are other **primary sources** of data, such as fieldwork.

Answering Eko's question

Eko wants to know if the day length on other planets is related to the year length. She knows that the Earth takes 24 hours to spin on its axis, and this is the length of its day. It takes 365 days to orbit the Sun. This is the length of its year. She finds a book and looks up the time it takes for different planets to spin once on their axis. She also looks up how long it takes them to orbit the Sun.

The book is a **secondary source** of data. Eko did not make the measurements herself. She used measurements that other people have made.

It is very important to check data, whether they are primary data or secondary data. This could mean repeating an experiment, or looking at more than one secondary source to check that the data are the same.

These are the data Eko found.

Planet	Time to spin once on its axis (hours)	Time to go once around the Sun (years)
Mercury	1416 (59 days)	0.24
Venus	2 137 440 (244 years)	0.69
Earth	24	1.00
Mars	25	1.88
Jupiter	10	11.86
Saturn	11	29.46
Uranus	17	84.01
Neptune	16	164.80

Eko looks at the data. She can use the data to answer her question.

Answering Yuda's question

Yuda wants to know about the seasons on other planets. He knows that we have seasons on Earth because the axis of the Earth is tilted. He looks on the Internet and finds a website about the planets. He writes down the angle at which the axes of the planets are tilted.

He checks on several other websites and finds that the data are the same on all the sites.

Data are not always presented in the form of a table. Yuda found the data on this diagram. He took the numbers from the diagram and made a table for his results. When you are using secondary data you may need to display the data in a different way, like a chart or a graph.

Planet	Angle of tilt (°)
Mercury	0
Venus	177.4
Earth	23.5
Mars	24.0
Jupiter	3.1
Saturn	26.7
Uranus	97.9
Neptune	28.8

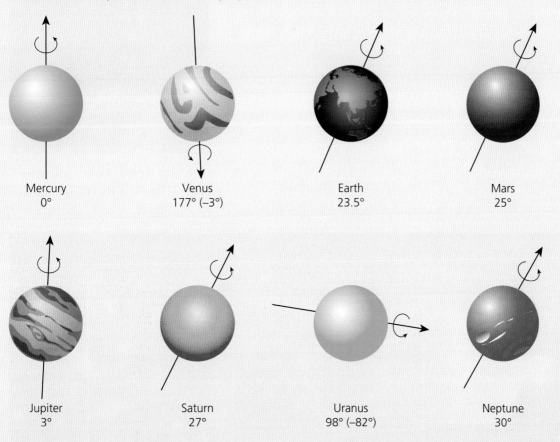

Mercury
0°

Venus
177° (−3°)

Earth
23.5°

Mars
25°

Jupiter
3°

Saturn
27°

Uranus
98° (−82°)

Neptune
30°

not to scale

Q

1 Explain the difference between primary and secondary sources.

2 Why is it important to look at more than one secondary source?

3 Look at Eko's data. Is there a link between day length and year length? Explain your answer.

4 Look at Yuda's data.

   a Which planet doesn't have seasons?

   b Which planets have seasons that would be similar to Earth's seasons?

   c On which planets would you have seasons but you would hardly notice them?

   d On which planet would the seasons be very different from the Earth's seasons?

!

- Experiments and fieldwork are primary sources of data.

- Secondary sources of information, such as a book or the internet, contain data that other people have collected.

- You can use primary and secondary data to answer questions.

The origin of the Universe

An astronomer looks at galaxies in the night sky. She works out that they are all moving away from the Earth. You might think that the Earth is special, or that it is the centre of the Universe, but that is not the case.

The Big Bang

Every galaxy is moving away from every other galaxy. Astronomers found evidence for this movement in the 1920s, when they used telescopes to measure how fast galaxies are moving. They also measured how far away they are. All the galaxies that they looked at were moving away from us. The further away they were, the faster they were moving.

Scientists used the evidence from their observations to develop an explanation about how the Universe began. This explanation is called the **Big Bang**. It is an idea that explains observations and predicts what might happen in the future.

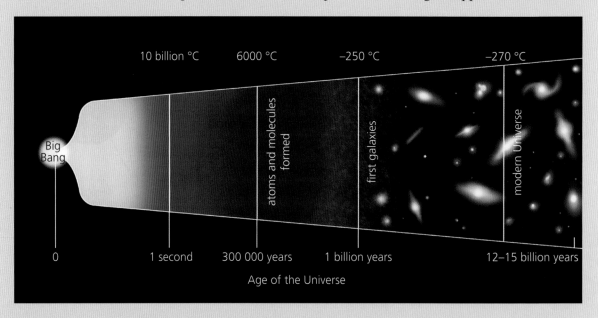

Age of the Universe

Most astronomers think that the Universe began about 14 billion years ago. The whole Universe **expanded** from something smaller than an **atom** and hotter than anything we can imagine.

In a fraction of a second the Universe grew to the size of a galaxy, and it has continued to expand ever since. As it expanded it cooled and energy changed to particles, making atoms and molecules of hydrogen and helium. Eventually there were stars, planets, Moons, and galaxies.

The galaxies are moving away from us because the space in between all galaxies is expanding. There is nothing special about our galaxy the Milky Way.

If we lived in the Andromeda galaxy it would still appear that all the other galaxies, including the Milky Way, were moving away from us.

A timeline for the Universe

There are many significant events in the history of the Universe. The formation of the Sun and the Solar System, the first plants and animals, the extinction of the dinosaurs, and the first humans are just some of those events. We can look at the history of the Universe on a timeline.

It is difficult to understand a long timescale like the history of the Universe. Sometimes it is easier to picture it with an **analogy**. If the Universe started 24 hours ago, then the Earth would have formed 9 hours ago, and the first animals would have appeared about an hour ago. Human beings would have existed for less than the time that it takes to blink your eyes.

The end of the Universe

No one knows what will happen to the Universe. It is possible that it will expand forever, or that gravity will pull the Universe back in again. To know if that is going to happen we would need to be able to measure the mass of all the objects in the Universe. This is not possible with the technology that we have at the moment.

Time (million years)	Event	
Now	today	
0.5	humans existed	
65	dinosaurs died out	
200	dinosaurs lived	
550	first land plants	
630	first animals with shells/ multicellular organisms	
3100	oldest rock	
3900	single-celled organisms	
4200	Earth cooled	
5000	Solar System formed	
13 700	Big Bang	

1 Look at the timeline for the Universe.

   a How old do scientists think the Universe is?

   b How old is the Solar System?

2 Why do scientists find it difficult to predict what will happen to the Universe?

3 Is it possible for humans to have seen a dinosaur? Explain your answer.

- Scientists think that the Universe began with a Big Bang about 14 billion years ago.

- The Universe is expanding and all the galaxies are moving away from each other.

1 Match each object to its correct definition. Write out the numbers and the letters.

Definitions:

1 a piece of dust or rock that burns up in the atmosphere

2 a piece of rock in an orbit between Mars and Jupiter

3 a meteor that has landed on the ground

4 a big, dirty snowball that orbits the Sun

Words:

A asteroid

B comet

C meteor

D meteorite [4]

2 Which of these statements explains why we have day and night?

a The Earth orbits the Sun once a day.

b The Sun orbits the Earth once a day.

c The Sun spins on its axis once a day.

d The Earth spins on its axis once a day. [1]

3 Look at the diagram of the Solar System.

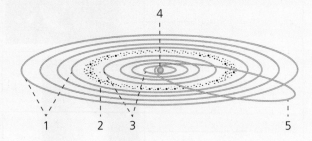

Write the number that shows where you find:

a the inner planets [1]

b the outer planets [1]

c the asteroid belt [1]

d a comet [1]

e a star. [1]

4 Here is a list of objects that you can see from the Earth, with the naked eye or with a telescope.

Sun comet Jupiter's moon star (not our Sun)

Moon planet Earth Venus Mercury

a Which objects can you see without using a telescope? [1]

b Explain why you cannot put the objects in order of size. [1]

c 'Mercury' could form a group with two other words on the list. Which two, and why? [1]

d 'Sun' could form a pair with one other word on the list. Which one, and why? [1]

5 Here is a diagram of the first four planets in the Solar System.

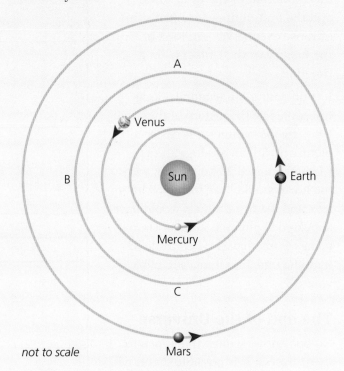

not to scale

a How many months later than its position shown above will the Earth be at A? [1]

b How many months later than its position shown above will the Earth be at B? [1]

c How many months later than its position shown above will the Earth be at C? [1]

d The distance between the Sun and Mercury is 60 million km. The distance between the Earth and the Sun is 150 million km. Work out the closest distance that Mercury gets to the Earth. [2]

e Use the diagram to explain why we see different stars at different times of the year. [1]

6 Copy and complete these sentences. You may use each word once, more than once, or not at all.

away from towards north south day month
year equator axis shorter longer

a We have seasons because the _____ of the Earth is tilted. The Earth orbits the Sun once every _____. [1]

b In summer in the southern hemisphere the _____ pole is tilted towards the Sun and the _____ is tilted away. This means that the days in the northern hemisphere are _____ and the days in the southern hemisphere are _____. [1]

c The energy from the Sun is spread out over a _____ area in the winter because the Earth is tilting _____ the Sun. [1]

7 When studying astronomy you cannot do experiments. Scientists make observations and develop explanations to account for the observations. Here is a list of observations that astronomers made thousands of years ago.

A The ground that we are standing on does not move.

B We see different stars in June and December.

C The Sun moves across the sky.

D Planets appear to wander across the sky.

E The stars do not appear to wander across the sky.

a Which of these observations fits with a geocentric model of the Universe? (There may be more than one.) [1]

b Which of these observations fits with a heliocentric model of the Universe? (There may be more than one.) [1]

c Which observation is explained more simply by the heliocentric model than by the geocentric model? [1]

d Describe one piece of evidence that a modern astronomer could show someone to demonstrate that the Earth is spinning. [1]

8 Below is the length of a year on each planet. Look at the distance of each planet from the Sun in the table on page 61.

Planet	Mercury	Venus	Earth	Mars	Saturn	Jupiter	Uranus	Neptune
Year length	88 days	225 days	365 days	687 days	11.8 years	29.4 years	84 years	165 years

a Describe the link between distance from the Sun and the length of one year. [1]

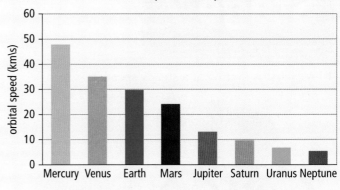

The speed of the planets

This chart shows the speed of the planets in their orbit.

b Describe the link between the distance from the Sun and the speed. [1]

c Why is one year on Uranus longer than one year on Earth? [1]

9 People have studied astronomy all over the world for many thousands of years. Different scientists came up with the same explanations for their observations.

a Give two reasons why a scientist working in Egypt might not have known about an explanation developed by a scientist in India. [1]

b Give one example from your studies in this topic where that happened. [1]

c Give two reasons why it is less likely that two scientists would separately develop the same explanation for the same evidence today. [2]

10 You can use primary and secondary data to develop explanations in science.

a Explain the difference between primary and secondary data. [1]

b An astronomer uses data from a science book to develop an explanation about the type of atmospheres on different planets. Are the data primary or secondary? [1]

c A student uses a table of numbers from a magazine to find out about the number of moons around all the planets. Are the data primary or secondary? [1]

d A teacher uses a telescope to make some observations of the craters on the Moon. Are the data primary or secondary? [1]

e Which data are likely to be more reliable? Explain your answer. [1]

1 A cyclist is cycling at a steady speed.

steady speed

a What are the forces acting on the cyclist to slow him down? [1]

b What can you say about the *size* of the drag forces acting on him, compared with the force that he applies to the pedals? [1]

c Explain your answer.
He crouches forward and makes himself more streamlined, but continues to pedal with the same force.

d What will happen to his speed? [1]

e Explain your answer. [1]

2 A student says: 'There is a force of gravity between any objects with mass. There is a force of gravity between me and my chair. The force of gravity between me and my chair makes it hard to get out of my chair!'

Which sentence or sentences are correct? [1]

3 A fish is swimming through water in a lake. Its tail provides the thrust so that it can move forward at a steady speed at the same height below the surface.

a What are the other forces acting on the fish? [1]

A fisherman uses a nylon fishing line to catch the fish.

He pulls the fish out of the water.

b The original length of the line is 60 cm. When the fish is on the line and out of the water, the length is 62 cm. What is the extension? [1]

c He notices that the fish feels a lot heavier out of the water than it did when it was in the water. Why? [1]

d The fisherman uses the same line to catch a fish with half the mass of the first one. How would the extension of the line be different? [1]

If the fishing line breaks it can be very dangerous because the end of the line can move very fast.

e What kind of energy does it have as it is moving through the air? [1]

f Where did the energy come from? [1]

4 A baby gorilla is playing.

He climbs up a tree and sits on a branch. He pulls some leaves off the tree and they fall to the ground. Then he jumps down from the branch. He sees his mother and they run towards each other.

a What type of energy did he gain as he climbed the tree? [1]

b Where did this energy come from? [1]

c He dropped a large leaf and a small leaf from the branch. Which one reached the ground first? Explain your answer. [1]

d Draw an energy transfer diagram that shows the changes from when he jumps off the tree to when he has landed on the ground. [1]

e His mother runs towards him at the same speed that he is running towards her. Who has more kinetic energy? Explain your answer. [2]

5 Here are some data about different types of fan.

Type of fan	Kinetic energy produced for each 200 J of electrical energy supplied (J)
A	70
B	100
C	80

a How much energy is 'wasted' by fan A? [1]

b In what form is the wasted energy? [1]

c Which law were you using to calculate the answer to part **a**? [1]

d List the fans in order of efficiency starting with the least efficient. Explain why you have chosen this order. [2]

6 A planet has been discovered orbiting a star that is not our Sun. The star (51 Pegasi) is about the same size and brightness as our Sun. Here is some information about the planet, which is called 51 Pegasi b.

	51 Pegasi b	Earth
Distance from the star (million km)	7.7	150
Time to orbit the star (days)	4	365
Time to spin once on its axis (days)	4	1
Tilt of the axis (°)	79	23.5

a Do you think the surface temperature of 51 Pegasi b is higher or lower than the temperature on Earth? Explain your answer. [1]

b How long is a day on 51 Pegasi b? [1]

c How long is a year on 51 Pegasi b? [1]

d How is 51 Pegasi b similar to our Moon? (*Hint:* look at your answers to parts **b** and **c**.) [1]

e Does 51 Pegasi b have seasons? Explain your answer. [1]

51 Pegasi is a star in the constellation of Pegasus. It is 51 light years away.

f What is a constellation? [1]

g Is a constellation the same as a galaxy? Explain your answer. [1]

7 Here are some statements about the geocentric and the heliocentric models of the Universe.

a Which the following statements are true and which are false? [2]

A The geocentric model could only work if you added smaller orbits to the orbits of planets.

B A heliocentric model explains the motion of planets more simply than the geocentric model.

C Nobody before Copernicus had thought of the heliocentric model.

D Copernicus built his heliocentric model using the ideas of others.

E The orbit of the Moon around the Earth was the evidence that Galileo used to convince people that the Earth orbited the Sun.

F The invention of the telescope was very important for the development of the heliocentric model.

G There were no astronomical observatories in India before Copernicus.

b Astronomers use telescopes to collect data. Are these primary or secondary data? [1]

c Scientists ask questions, and then use evidence to develop explanations and test them. Think about how the question/evidence/explanation system fits in with the development of the heliocentric model.

Here are three sentences.

A The planets seem to wander across the sky, and there are moons around other planets.

B Do all objects in the Solar System orbit the Sun?

C Planets orbit the Sun and moons orbit planets.

   i Which is the question, A, B, or C? [1]

   ii Which is the evidence, A, B, or C? [1]

   iii Which is the explanation, A, B, or C? [1]

Speed

↑ Speed limits are described in km/h in India.

How fast?

Bicycles, cars, and airplanes all travel at different **speeds**. Speed is a measure of how fast something is moving. It is the distance moved per second or per hour. The unit of speed is **metres per second** (m/s) or **kilometres per hour** (km/h). A very fast car can drive at over 300 km/h, but these girls are probably walking at 5 km/h which is about 1 m/s.

You may have heard people describe the speed of a cricket or tennis ball as moving at 80 miles per hour, which would be written down as 80 mph. In the UK and the USA speeds are usually measured in mph rather than km/h.

How to calculate speed

To find the speed of a moving object, you measure the time it takes to travel between two points a set distance apart. You work out the speed from the distance divided by the time:

$$\text{Speed} = \frac{\text{distance}}{\text{time}}$$

A long-distance runner runs part of his race at a **steady speed**. It takes him 20 seconds to run 100 m.

$$\begin{aligned}\text{Speed} &= \frac{\text{distance}}{\text{time}}\\ &= \frac{100 \text{ m}}{20 \text{ s}}\\ &= 5 \text{ m/s}\end{aligned}$$

When you are doing a calculation it is helpful to write it out like this. If you put the units of distance and time in the equation, you will have the correct units for the speed.

Average speed

In May 2010 Dilip Donde became the first Indian to sail around the world single handed. He had to sail through very bad storms near Antarctica, but arrived back safely in Mumbai harbour.

He sailed 34 560 km in 276 days, which is 6624 hours. His speed varied depending on the strength of the wind. In good sailing weather with lots of wind he went faster. If there was a storm, wind speeds could reach hurricane force (over 100 km/h), but sometimes he was stuck for days with very little wind.

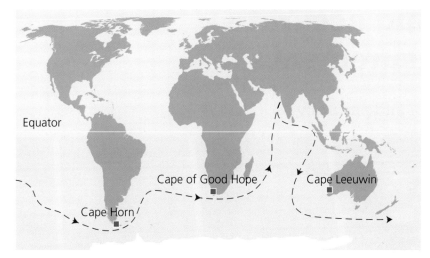

↑ Many round-the-world sailors take this route.

We can work out what speed he did on **average** over the whole journey by dividing the *total distance* by the *total time* that it took. This **average speed** makes it easier to compare how fast different people, or boats, or cars travel.

$$\text{Average speed} = \frac{\text{total distance}}{\text{total time}}$$

$$= \frac{34\,560 \text{ km}}{6624 \text{ h}}$$

$$= 5.2 \text{ km/h}$$

	Speed (m/s)	Speed (km/h)
walking quickly	1.7	6.1
sprinting	10	36
typical driving speed limit	14	50
category 1 hurricane	33	119
plane cruising speed	255	917

The table shows some typical average speeds.

During a journey the speed of a car will change as the car speeds up and slows down. The speed that the driver sees on the speedometer is the speed at that moment in time. That speed is worked out from the size of the wheels on the car. It is not the average speed.

Q

1 David drives 150 km in 2 hours. Calculate his average speed in km/h.

2 Fatima runs a 200 m race in 40 seconds. What is her average speed?

3 Preeti runs 100 m in 19 seconds. Nikita runs the same distance in 15 seconds. Who is faster?

4 Look at the statements below. Decide whether each one is true or false. Then use the equation for speed to work out how fast all the objects are moving. Were you correct?

  a A car that travels 100 km in 2 hours is going faster than a vehicle that travels the same distance in 3 hours.

  b A girl walking 10 m in 4 seconds is slower than a boy who walks the same distance in 3 seconds.

  c A motorbike that drives 50 km in 30 minutes is going faster than a car that travels 100 km in 45 minutes. (Remember: 30 minutes = 0.5 hours; 45 minutes = 0.75 hours.)

5 A student says that a speed of 50 m/s is faster than a speed of 140 km/h because the number is bigger. What would you say to them?

- Speed is distance divided by time.
- Speed is measured in m/s or km/h.
- To calculate average speed you need to know the total distance and the total time.
- Your speed during a race or a journey can change.

Taking accurate measurements

Precision and accuracy

⬆ Usain Bolt was the fastest man in the 2012 Olympics.

Usain Bolt won an Olympic gold medal in 2012 by running 100 metres in a record time of 9.63 seconds. The world's second fastest man in 2012, Yohan Blake, ran it in 9.75 seconds. Justin Gatlin, who won the bronze medal, took 9.79 seconds. When differences are so small, athletes have to be timed as accurately and precisely as possible.

Precision is shown by the number of **significant figures**. Without 3 significant figures it would be impossible to say whether Yohan Blake or Justin Gatlin took the silver medal. Both are 9.8 seconds if you use 2 significant figures.

Accuracy tells you how correct a measurement is. A measurement can be very precise, but not be accurate. If Usain Bolt's time had been recorded as 9.73 seconds it would have the same precision, but would have been very inaccurate.

Reaction time

For a long time the 100 m race was timed with a stopwatch.

In the 1970s automatic timing was introduced. The starter's gun triggers a timing clock, and the winner breaks a light beam when their body crosses the line. An optical device senses that the light beam has been cut and stops the clock. In the London 2012 Olympics, times for the 100 m were measured to a precision of 4 significant figures. This meant they could work out who won even if the times to 3 significant figures were the same.

When automatic timing was first introduced, the measured times for the 100 m got longer! When the judge used to use a stopwatch there was a short delay between hearing the gun and starting the stopwatch. This delay is called the **reaction time**. It is the time that it takes the brain to process information. It is typically about 0.2 seconds. This delay meant that the times were not accurate.

The judge then stood by the finish line to finish the race. They could anticipate when the athlete would cross the line, so there was not the same reaction time when they stopped the stopwatch. The times measured were shorter than they should be by about 0.2 seconds.

Measuring time in the laboratory

It is important to think about reaction time if you are measuring time in an experiment.

If you are timing something over a long period of time, then your reaction time will not be so important. If you are timing something that happens very quickly then it will have a big effect.

Dev is doing an experiment about air resistance. He is dropping a ball from a metre off the ground. He can anticipate when the falling ball will land, so again the reaction time makes the reading inaccurate.

The reading on his stopwatch is 0.65 seconds. If Dev's reaction time is 0.2 seconds, then that makes a big difference to the time that he needs to record. The accurate reading could be 0.45 seconds!

It is very difficult to measure times that are this short with a stopwatch. Dev could improve the accuracy of his experiment by using **timing gates**. These gates work like the finish of the 100 m race. As the ball passes through the top gate, it breaks a light beam and starts the timer. When it passes through the bottom gate, it breaks another beam that stops the timer. These timers can record times that are accurate and precise to 3 significant figures.

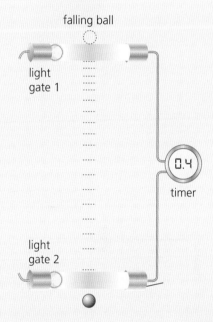

falling ball

light gate 1

light gate 2

timer

Q

1 Why do you get a more accurate measurement of the time that it takes a ball to fall using timing gates rather than a stopwatch?

2 Why is automatic timing more important for a sprint race than a marathon?

3 Kyra and Pari are timing how long it takes for 1 g of sugar to dissolve in hot water. Kyra records a time of 24.3 seconds. Pari records a time of 24 seconds. Whose reading is more precise? Why?

4 When sprinters start a race, sensors in their starting blocks can register a false start. They detect a change in pressure against the block as the athlete starts. If the time difference between the starting gun going off and the change in pressure is too short, a false start is registered.

   a Which of these times could be the time difference that gives a false start: less than 0.2 seconds, 0.2 seconds, more than 0.2 seconds?

   b Explain your answer.

!

- Accuracy is how close a measurement is to its true value.

- Precision is the number of significant figures to which a value is given.

- Your reaction time is about 0.2 seconds.

- The reaction time is the time that it takes the brain to process information.

Distance–time graphs

What is a distance–time graph?

A **distance–time graph** is a useful way of showing how something is moving.

You can collect data about a moving object by measuring the distance it has travelled from its start point each second. Then you can plot a distance–time graph from those data.

A cyclist is photographed every 2 seconds as he cycles along a track. The lines on the track show the distances. These data are displayed in a table and on a graph.

Time (s)	Distance (m)
0	0
2	12
4	24
6	36
8	48
10	60

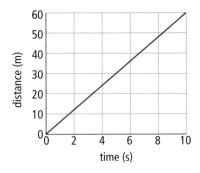

Telling a story

In each second the cyclist moves the same distance – his speed does not change. This means the graph is a straight line.

The slope, or **gradient**, of a distance–time graph tells you the *speed*. If the line is steep the object is moving fast. If it is not very steep then the object is moving more slowly.

In the graph on the right, both objects are moving at a steady speed but line A shows a fast speed and line B shows a slower speed.

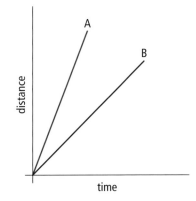

The graph on the left also tells a story, but not a lot is happening! The line is flat, so the slope is zero. This tells you that the speed is zero – the object isn't moving.

A journey to school

Rana walks to school. She stops to chat, and changes speed along the way.

At the top of the next page is a distance–time graph of Rana's journey to school.

Sections B and D of this graph are flat. This tells you that Rana had stopped. In sections A, C, and E, the graph is a straight line that is sloped. She is moving at steady speed.

Rana's graph suddenly changes between flat sections and sloped sections. But usually when we are walking we don't suddenly change speed – our speed changes continuously. The second distance–time graph is more realistic.

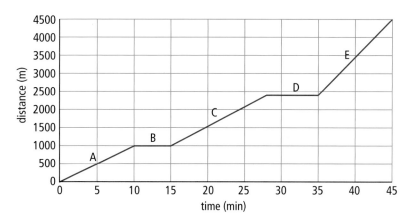

Working out the speed from a distance–time graph

You can calculate the speed of a moving object from its distance–time graph. For example, in section A of the graph Rana walks 1000 m in 10 minutes. Ten minutes converted to seconds is 600 seconds.

$$\text{Average speed} = \frac{\text{distance}}{\text{time}}$$
$$= \frac{1000 \text{ m}}{600 \text{ s}}$$
$$= 1.7 \text{ m/s}$$

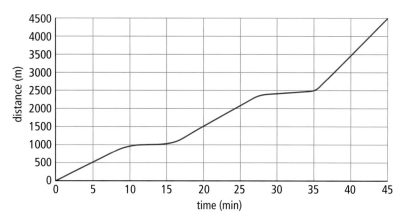

Changing speed

When you drop a ball its speed gets bigger as it falls. A distance–time graph for the ball will not be a straight line because the speed is not constant. It is a curve and the slope of the curve is increasing. This shows that the speed is increasing.

A distance–time graph of something that is slowing down would also be a curve, but the line would become less and less steep. This would show that the speed is decreasing.

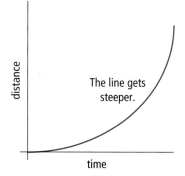

The line gets steeper.

1 In which section of her journey to school is Rana moving fastest? How do you know?

2 Three students cycle 4.5 km to school. Their distance–time graph is shown here.

  a Who arrived first?

  b Who stopped? How do you know?

  c Who was slowest at the start?

  d Who was slowest at the end?

  e What was the average speed of the person who arrived first?

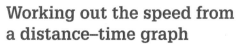

3 Calculate Rana's speed for sections C and E of the graph at the top of the page.

- A distance–time graph shows how the distance moved changes with time.
- A straight, upward-sloping line means a steady speed.
- If the line is horizontal, then the object is not moving.
- If the line is steeper, the object is moving faster.

Acceleration and speed–time graphs

Changing speed

When the speed of something changes, we say that it is **accelerating**. The speed of a ball can change quickly. Acceleration tells you how much the speed changes in one second.

Calculating acceleration

To calculate **acceleration** you need to know:

- the speed at the start

- the speed at the end

- the time it took for the speed to change.

$$\text{Acceleration} = \frac{\text{change in speed}}{\text{time}}$$

or

$$\text{Acceleration} = \frac{\text{final speed} - \text{starting speed}}{\text{time}}$$

We measure it in m/s *per second* or metres per second squared (m/s^2).

Sinita is using her scooter to get to work. She sets off from her house and accelerates from 0 to 20 m/s in 5 seconds.

$$\text{Her acceleration} = \frac{\text{final speed} - \text{starting speed}}{\text{time}}$$

$$= \frac{20 \text{ m/s} - 0 \text{ m/s}}{5 \text{ s}}$$

$$= 4 \text{ m/s}^2$$

Slowing down: decelerating

If the final speed is lower than the starting speed, it means that the object is slowing down or **decelerating**. The acceleration is negative.

Sinita sees a friend as she rides. She slows down from 20 m/s to 10 m/s in 2 seconds.

$$\text{Her acceleration} = \frac{10 \text{ m/s} - 20 \text{ m/s}}{2 \text{ s}} = \frac{-10 \text{ m/s}}{2 \text{ s}}$$

$$= -5 \text{ m/s}^2$$

We say that her **deceleration** is 5 m/s^2. This is the same as an acceleration of –5 m/s^2.

Speed–time graphs

You have seen that a distance–time graph shows how the distance changes with time. You can use it to find the average speed. A **speed–time graph** shows how speed changes. You can use it to find the acceleration by reading numbers from the axes of the graph.

In this graph the speed is increasing by 10 m/s every second. The acceleration is 10 m/s^2. It is **constant** – the line is straight and the gradient, or slope, is constant. If the acceleration was bigger then the line would be steeper.

If the speed does not change, then the line is horizontal. The acceleration is zero.

In the graph below the speed is a constant 5 m/s (zero acceleration) for 30 seconds. Then it drops to zero. The object has stopped.

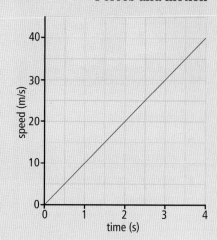

Terminal velocity

A cyclist is riding a race. He accelerates at the start and gets faster and faster. As he does so the force of air resistance increases until eventually the it balance the forward force. He cannot go any faster – he has reached **terminal velocity**. A speed–time graph tells the story of this journey. At first the line is steep, showing that his speed is increasing very quickly. The line gets less and less steep as his speed changes less quickly. This does not mean that he is slowing down. It means that it takes longer for his speed to increase. Eventually he travels at a steady, terminal velocity. The line is horizontal because his speed is no longer changing.

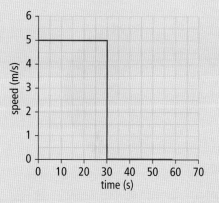

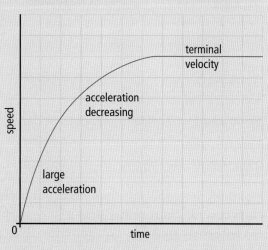

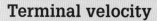

Q

1 Explain the difference between speed, acceleration, and deceleration.

2 Copy these sentences, choosing the correct words from each pair.
The slope of a speed–time graph tells you the **acceleration/distance**. If the line is **steep/horizontal** the speed is not changing. This **is/is not** the same as a speed of zero.

3 A hockey ball accelerates from 0 m/s to 25 m/s in 0.05 seconds. What is the acceleration of the ball?

4 A football decelerates from 10 m/s to 0 m/s in 0.1 seconds. What is the acceleration of the ball? What is the deceleration of the ball?

5 Look at this speed–time graph. Car A is accelerating at a constant rate of 12 m/s^2.

a Which has the bigger acceleration, car A or car B? How do you know?

b Use the information on the graph to confirm this.

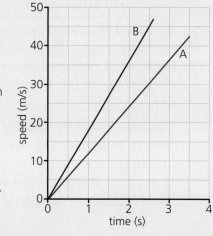

- Acceleration is a change in speed divided by time.

- If the object is slowing down, the acceleration is negative. It is decelerating.

- A straight, sloping line on a speed–time graph shows constant acceleration.

- A horizontal line on a speed–time graph shows constant speed.

Presenting results in tables and graphs

Objectives

- Present results in tables, charts, and graphs
- Explain what a continuous variable is

Formula 1 racing

The Force India Formula 1 team has been racing since 2007 on race tracks all over the world, including India, Korea, and Malaysia. Formula 1 cars are designed to reach speeds far greater than ordinary cars.

In the 2012 season there were 20 races in the World Championship. The table shows the number of races won by each driver. This table arranges the data alphabetically by surname.

Number of races won

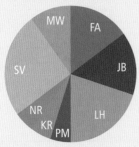

⬆ Pie chart. It has 20 equal parts, one per race. The slice for each race is 360° / 20 = 18°.

Winning driver	Number of races won
Fernando Alonso	3
Jenson Button	3
Lewis Hamilton	4
Pastor Maldonado	1
Kimi Raikkonen	1
Nico Rosberg	1
Sebastian Vettel	5
Mark Webber	2

We could show this information in a **bar chart** or a **pie chart**. We use a bar chart or a pie chart rather than a line graph because the name of the driver is a **categoric** variable. Categoric variables are words not numbers. The number of races is a **discrete** variable. You can only have whole numbers of races. If one of the variables in your table is a discrete variable, then you will need to plot a bar chart. A variable that can take any number, like time or distance, is **continuous**. The data in both columns must vary continuously to plot a line graph.

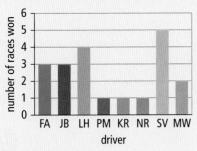

⬆ Bar chart.

Drawing line graphs: speed

The speed of a racing car changes as it goes round the track – the car accelerates and decelerates. The speed of a racing car is a continuous variable because it varies all the time.

This table shows the speed of a car during the first 20 seconds of a race. Both the table of data and the graph show that the speed is changing. It is much easier to see *how* the speed is changing by looking at the graph.

Time (s)	Speed (km/h)
0	0
2	100
4	195
6	270
8	310
10	320
12	320
14	320
16	300
18	280
20	300

- When the line slopes upwards, the car is accelerating.
- When the line slopes downwards, the car is decelerating.
- When the line is flat, the car's speed is constant.

The line is smooth because the speed changes continuously, not suddenly.

Drawing line graphs: distance

This table shows the *distance* that the car travelled in the first 20 seconds of the same race. We can show these data on a line graph too. Like speed, distance changes continuously.

Time (s)	Distance (m)
0	0
2	28
4	80
6	170
8	330
10	510
12	680
14	860
16	1030
18	1200
20	1360

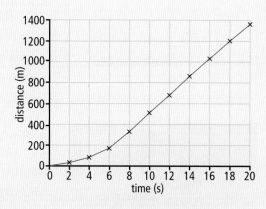

Telling a story

On a distance–time graph for a car:

- when the line is horizontal the car is stationary
- a sloping line means that the car is moving at a steady speed
- when the line curves up the car is accelerating.

On a speed–time graph for a car:

- when the line is horizontal the car is moving at a steady speed
- a sloping line means that the car is accelerating at a steady rate.

You can calculate the acceleration from a speed–time graph by reading the values of speed and time from the axes.

Q

1 Why are the number of races that each driver has won shown on a bar chart, but the speed and time of a car are shown on a line graph?

2 Look at the speed–time graph. Between which times was:

  a the speed constant?

  b the car decelerating?

  c the car accelerating?

3 Look at the distance–time graph.

  a How far did the car travel in the first 9 seconds?

  b Why is the distance–time graph not horizontal between 10 and 14 seconds like the speed–time graph?

4 The car stops in the pits for 10 seconds. How would this be represented on:

  a the distance–time graph

  b the speed–time graph?

5 At the start the Formula 1 car accelerated from 0 km/h to 200 km/h in 4 seconds. 200 km/h is the same as 55 m/s.

  a Calculate the acceleration.

  b Later it slowed down from 320 km/h to 280 km/h in 4 seconds. 280 km/h is the same as 78 m/s and 320 km/h is the same as 89 m/s. Calculate the average acceleration. What is the average deceleration?

- Variables that can have any numerical value are continuous.
- Pie charts and bar charts are used to show data when a variable is not continuous.
- Line graphs are used when both variables are continuous.

Asking scientific questions

Objectives

- Recognise scientific questions
- Describe how explanations are developed

Starting and stopping

Moving objects are slowed down by friction, air resistance, and water resistance. It takes a force to start something moving that is stationary.

Once something is moving, it also takes a force to keep it moving against these forces. To cycle at a steady speed, the boy pushes the pedals around. If he stops pushing on the pedals he will slow down and stop.

Asking questions about motion

Galileo was a scientist who lived over 400 years ago. He was interested in experiments about motion. He asked this question:

He wondered what would happen if there was no friction. It was impossible for him to do experiments without friction in a laboratory. There is always some friction between surfaces, even if they are really smooth.

Instead of doing an experiment with equipment, Galileo did a 'thought experiment'. This can be very useful to work out what might happen in a situation where it is not possible to carry out the experiment.

Galileo's 'thought experiment'

Galileo knew that something rolling down a slope speeds up.

He imagined a ball rolling down a slope, along a flat surface, and up another slope that was just as steep as the first. He knew that friction would slow the ball down. But what would happen if there was no friction? He realised that the ball would reach the same height that it started from.

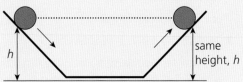

same height, *h*

Next he imagined the ball rolling up a longer slope. What would happen then? The ball would continue until it reached the same height again, he thought.

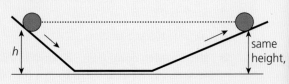

same height,

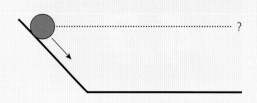

Finally he made the final slope longer and longer … so long that it was completely flat. What would happen then? Galileo realised that the ball would keep rolling at a steady speed forever! There is no force slowing it down. He had shown with his thought experiment that you do *not* need a force to keep something moving. This is an example of someone using **creative thinking** to solve a problem in physics.

What happened next?

Galileo wrote about his ideas in a book about motion in 1590. The idea that you did not need a force to keep something moving wasn't very popular. People believed that you did need a force to keep something moving, because that is what they saw in the world around them. They didn't realise that in most situations you need to apply a force to cancel out friction or air resistance. Without friction and air resistance this force is not needed.

It took a long time for his idea to be accepted.

Newton's first law and inertia

Galileo had shown in his thought experiment that without friction a ball does not need a force to keep it moving at a steady speed.

Sir Isaac Newton was a scientist who thought about how things move. He wrote his laws of motion in a book in 1687, about 100 years after Galileo wrote his book. He developed Galileo's ideas to work out three laws of motion. These laws explain why things are stationary, speed up, slow down, or move with a steady speed.

This is Newton's first law of motion:

"Every object in a state of uniform motion tends to remain in that state of motion unless an external force is applied to it."

This means that something that is stationary will remain stationary. Something that is moving in a straight line at a steady speed will continue to move in a straight line at a steady speed.

There is a resistance to motion changing – this is what Galileo had called **inertia**. It is obvious when the object that you are trying to move has a very large mass.

Newton's second law describes how resultant forces change the speed or direction of motion. Newton's third law describes how forces always come in pairs.

⬆ A massive object has a large amount of inertia. It takes a big force to make it accelerate.

Q

1 Imagine that there is no friction for a day. Make a list of things that it would not be possible for you do. Which things would still be possible?

2 A boat on a river is moving with a steady speed. The engine is running.

   a Explain why you need to keep the engine running to move at a steady speed.

   b What would happen if the engine was turned off?

3 Do you think it would make a difference if the slopes in Galileo's experiments had been curved rather than straight? Explain your answer.

- Some questions can't be tested easily using equipment.
- You can use your imagination to do thought experiments.
- Sometimes it takes a long time for an idea to be accepted.

1 Which of these speed calculations is correct?

 A A car that travels 100 km in 2 hours has an average speed of 50 km/h.

 B A fish that swims 20 m in 5 seconds has an average speed of 5 m/s.

 C A rocket that travels 100 km in 30 minutes has an average speed of 200 km/h.

 D A ball that rolls 2 m in 5 seconds has an average speed of 0.4 m/s.

[2]

2 Here are some units: km, m/s, km/s, h, s, m.

 a Which are units of speed? [1]

 b Which are units of distance? [1]

 c Which are units of time? [1]

3 A girl is cycling at a steady speed.

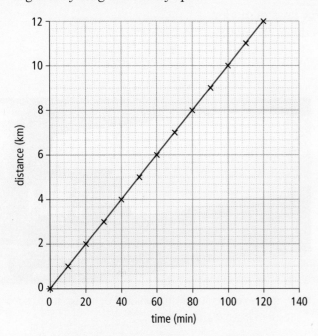

 a How can you tell from the graph that the speed is steady? [1]

 b How fast is she travelling? [2]

4 Choose the *best* definition of average speed.

 a how far you travel per second

 b the total distance that you travel divided by the time for the journey

 c how far you travel in one hour

 d the total distance that you travel divided by the total time that it takes [1]

5 Rafal went to visit his grandmother in an aeroplane, a distance of 1440 km.

 a The journey took 2 hours. How fast did the plane travel? [2]

 b When the plane took off it had to speed up. What is the scientific word for speeding up? [1]

 c On the way home the plane had to fly into the wind. What effect would that have on the average speed? [2]

6 Jameel was watching his favourite football team play at their stadium.

 a He saw the striker run the length of the field, which is 110 m or 0.11 km, in 10 seconds. How fast did the striker run? [2]

 b Jameel walked home, a distance of 3 km in 30 minutes. How fast did he walk? [2]

7 In experiments it is important to make measurements that are both precise and accurate. Which of these measurements of time is the most precise?

 5.1 seconds   5.14 seconds   5 seconds

[1]

8 Aadi went to visit Dev. The graph shows how Aadi's distance from home varies with time.

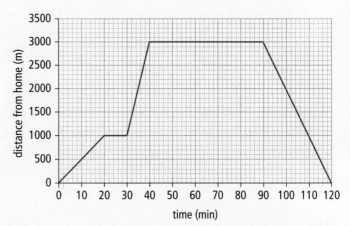

 a For how long does Aadi stop on the way to Dev's? [1]

 b How far is it from Aadi's home to Dev's home? [1]

 c How long does Aadi spend at Dev's? [1]

 d On which part of the journey does Aadi travel the fastest? [1]

e How fast does Aadi walk on the way back from Dev's home? [1]

9 A tortoise travelled across a path 1 m wide in 10 minutes. It then rested for 2 minutes and went across a 2 m wide piece of grass in the next 8 minutes.

a How fast was it travelling when it crossed the path? [2]

b How fast was it travelling when it crossed the grass? [2]

c What was the average speed for the whole journey? [2]

10 Nadia is looking at some information about scooters in a showroom. There is information about the colour, the top speed, the engine size, the manufacturer, and the cost.

She makes tables to show the data.

Colour	Engine size
Red	50
Blue	125

⬆ Table A

Cost	Engine size
12 000	50
14 000	125

⬆ Table B

Manufacturer	Cost
Speedyscooter	12 000
Scooterland	14 000

⬆ Table C

Top speed	Cost
80	12 000
100	14 000

⬆ Table D

a Which two tables of data can be used to plot line graphs? [1]

b Which two tables of data can be used to plot bar charts or pie charts? [1]

c What has Nadia missed out of her tables? [1]

d Which of the variables are categoric? [1]

e Name one of the continuous variables. [1]

11 Here is a speed–time graph for a car.

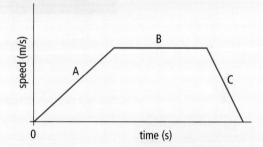

Are these statements true or false?

a The car is stationary in section B. [1]

b The car is slowing down in section C. [1]

c The car is moving at a steady speed in section A. [1]

d The car is moving at a steady speed in section B. [1]

e In section A the car is moving with steady speed that is slower than the steady speed in section C. [1]

12 A student records the time that it takes different types of ball to fall through a distance of 1.5 m. He writes down these results.

tennis ball 0.53 s, football 0.68 s, table tennis ball 0.72 s

a The student should have used a table to display his results. Draw the table and put the results in the table. [2]

b The student used timing gates to measure the time. How do timing gates work? [1]

c Calculate the average speed for each ball. [3]

d Which graph would you draw to show the average speed of the balls? Explain your answer. [2]

Sound, vibrations, and energy transfer

Making sounds

Your vocal cords, a drumskin, and a loudspeaker cone are all examples of things that **vibrate** to produce **sound**. Sound transfers energy from a **source** such as a drum to a **detector** such as your ear.

Sound needs a material, or **medium**, to travel through, such as air, water, or walls. The vibrating source produces a disturbance in the medium. This disturbance travels from the source to your ear, and that is when you hear the sound.

You can hear sounds when you are underwater. Animals such as dolphins and whales communicate over very large distances by making and hearing sounds.

If someone is talking loudly in the next room you will probably hear them, because sound travels through solid materials like walls. Sound cannot travel through empty space. A space that is completely empty of everything, including air, is a **vacuum**.

What is a sound wave?

If you put small polystyrene balls on a loudspeaker cone, they will bounce up and down, showing that the cone is vibrating. The cone moves out, then back, then out again.

The vibrating loudspeaker cone makes the air molecules next to it move backwards and forwards. This disturbance, or **vibration**, is what we call a **sound wave**. The layer of air next to the loudspeaker vibrates, which makes the next layer of air molecules vibrate, and so on. In this way a sound wave moves through the air.

The air itself does not move away from the cone. The sound wave transfers energy through the motion of the air molecules, and the wave moves from the loudspeaker out to your ear.

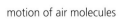

motion of air molecules   sound wave moves this way

You cannot see the air molecules moving, but you can make a model of a sound wave using a Slinky spring. If you hold the end and move your hand forwards and backwards, the coils move closer together then further apart as the wave moves along the Slinky. The individual coils do not travel to the end of the Slinky, but the wave does.

This is what happens to the air molecules in a sound wave. Where the air molecules are close together it is called a **compression**. Where they are further apart it is called a **rarefaction**.

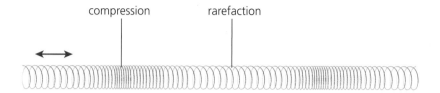

The speed of sound

Sound waves travel at different speeds in solids, liquids, and gases. They travel fastest in solids and slowest in gases. This is because the particles in a solid are closer together than they are in a gas, so the vibration is passed on more quickly.

Material	Speed of sound (m/s)
air at sea level	330
water	1500
metal	5000

Longitudinal and transverse waves

You can make two types of **wave** using a Slinky.

Waves that are made by moving your hand in and out are called **longitudinal** waves. The coils of the slinky move to and fro in the *same* direction as the wave is travelling. Sound waves and some types of earthquake wave are longitudinal waves.

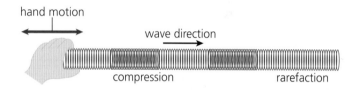
longitudinal wave

You can make a different type of wave by moving your hand from side to side instead of in and out. These waves are called **transverse** waves. The coils of the slinky are moving *at right angles* (90°) to the direction of travel of the wave. Stadium (or Mexican) waves made by fans at football matches are an example of a transverse wave. The people move up and down as the 'wave' moves along the stands.

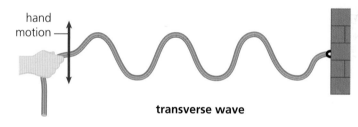
transverse wave

Q

1 Your vocal cords and a loudspeaker produce sound. Name three other sources of sounds waves.

2 Explain why sound cannot travel through a vacuum.

3 Does sound travel faster or slower in air compared with water? Explain your answer.

4 How is sound produced in a guitar?

!

- Sound waves are produced by vibrating objects.
- Vibrations are transferred through solids, liquids, and gases by particles vibrating.
- Sound travels fastest in solids and slowest in gases. It cannot travel through a vacuum.

Detecting sounds

Your **ear** and a **microphone** both detect sounds.

Objectives

- Describe how the ear detects sound
- Explain how your hearing can be damaged
- Describe how a microphone works

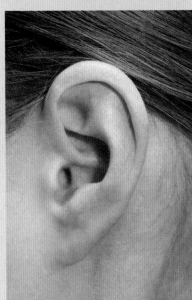

The human ear

You can only see a small part of your ear when you look at it in the mirror. Your ear has parts *inside* that detect sounds and send signals to your brain.

Having two ears helps you work out where sounds are coming from. Sounds reach your ears at different times and this enables you to locate the source of the sound.

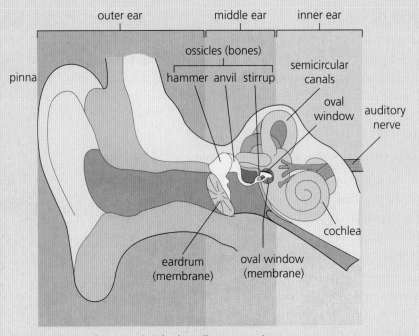

↑ The ear has parts inside that allow you to hear.

The outer and middle ear

The **outer ear** is made up of the **pinna**, **auditory canal**, and **eardrum**. The outer ear gathers the sound wave and directs it to the eardrum. Once there it makes the eardrum vibrate, and this in turn makes the **ossicles** vibrate.

The ossicles make up the **middle ear**. They are the smallest bones in your body. They pass the vibration on to the **oval window**, and then the **inner ear**.

Inside the inner ear

The inner ear is made up of two parts: the **semicircular canals** and the **cochlea**. The semicircular canals help you to balance.

The cochlea is shaped like a snail shell. It contains fluid and also sound-detecting cells with hairs on them. When the oval window vibrates, it transmits the vibration to the fluid. The vibrating fluid makes these hairs vibrate. The cells release chemicals which produce a signal that travels down the auditory nerve to your brain. Your brain processes the signal and you hear the sound. You can find out more about specialised cells like sound-detecting cells in your *Complete Biology for Cambridge Secondary 1* Student book

Hearing loss

Your ear contains delicate structures including the tiny ossicle bones and thin membranes of the oval window and eardrum. There are many ways that your hearing can be affected if these structures are damaged.

- Loud sounds can **perforate** (make a hole in) your eardrum. The hole will usually heal in a few weeks or months.

- Many people get wax in their ears, and this can affect your hearing. It is very easy to remove excess wax.

- If you have an ear infection, fluid can be produced around the small bones. This interferes with the transfer of the sound wave from the outer ear to the inner ear.

- Head injuries can affect the auditory nerve which will affect your ability to hear properly.

Older people do not hear high-pitched sounds as well as younger people. This is because there are gradual changes to your inner ear as you get older. Some people wear hearing aids to improve their hearing.

⬆ Loud sounds, like the music at concerts, can damage the eardrum.

⬆ As you get older your hearing can change.

Microphones and loudspeakers

A **microphone** is a type of **transducer** – it converts the energy in a sound wave into an **electrical signal**. The human ear is the body's microphone.

The microphone contains a diaphragm, which is a flexible plate. When the sound wave hits the diaphragm it makes it move backwards and forwards, like the eardrum in your ear. This movement produces an electrical signal that can be amplified and sent to a loudspeaker, or recorded and stored.

A **loudspeaker** is a another type of transducer – it converts the electrical signal into a sound wave. The electrical signal makes the cone move in and out, which makes the air molecules move backwards and forwards to make a sound wave.

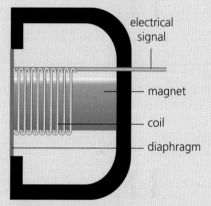

⬆ Like your ear, a microphone produces an electrical signal from a sound wave.

1 In what ways is your ear like a microphone?

2 Which part of a microphone is like the eardrum?

3 Copy and complete the table.

Part of the ear	What it does
pinna	
eardrum	
	passes the vibration from the eardrum to the cochlea
cochlea	

4 Why would it be dangerous to put a sharp object into your ear?

5 A student says 'a microphone is a bit like a loudspeaker in reverse'. Do you agree? Explain your answer.

- The membranes and bones of your middle ear transfer the vibration of a sound wave from your outer to your inner ear.

- Your inner ear converts the energy in the sound wave into an electrical signal that goes to your brain.

- A microphone produces an electrical signal that allows us to make recordings of sound.

Loudness and the decibel scale

Noise

⬆ We like listening to some sounds…   ⬆ …but not to others.

Our ears are very sensitive to sounds. Most people can hear very quiet sounds and also very, very loud ones. There are lots of sounds that we like, such as music and the voices of our family or friends. Sounds that we do not want to hear are called **noise**.

Sound levels

Noises can be very loud, or can go on for a long time, or both. The **intensity** of a sound is measured with a **sound level meter** on a scale called the **decibel** (**dB**) **scale**. A sound with a high intensity will sound louder.

Sound level (dB)	Activity	What this sounds like
0	threshold of hearing	cannot be heard
10	soft whisper	can just be heard
20	leaves rustling	very quiet
30		
40	talking quietly up to normal speech	quiet
50		
60		
70	noisy street	difficult to hear someone talking
80	traffic at a junction	annoying
90	truck chainsaw	can damage hearing after 8 hours per day
100	jet taking off 500 m away	can damage hearing after 2 hours per day
110	thunder	can damage hearing after 1.5 hours per day
120	vuvuzela or nadhaswaram	
130	road drill	painful
140	jet engine 100 m away	loudest recommended level with ear protection
194	loudest sound that it is possible to make	

One of the world's loudest musical instruments is the nadaswaram from India. The vuvuzela is another loud instrument – in the Football World Cup in South Africa the crowd used horns that could reach a level of 120 dB. The loudness of the loudest possible *continuous* sound is 194 dB. Explosions, earthquakes, or meteor impacts can produce much louder sounds.

The decibel scale is not like a ruler. A 50 dB sound is 100 times as intense as a 30 dB sound. Every increase of 10 dB increases the intensity 10 times.

Loudness and hearing

Loud sounds can damage your hearing. The risk depends not only on the number of decibels but also on the length of time you are exposed to the sound.

When scientists talk about the **risk** of an activity, they consider the **probability** that something bad will happen, and the **consequence** it if did. Some activities have very serious consequences, like an aeroplane crashing, but the probability of them happening is very low.

We often take risks when the consequences do not seem serious. Listening to music or going to concerts may not seem risky, but there is a probability that your hearing will be damaged by loud music. That probability gets bigger the longer you listen to the music.

Reducing risk

Some people have to hear loud noises because of their jobs. They need to **shield** their ears from loud sounds by wearing ear protection.

There are three main ways of reducing the risk from noise if you cannot change the sound level:

- shielding – putting something between the source of the sound and your ears, such as **ear defenders**
- increasing the distance – moving away from the source of the sound
- reducing the time that you spend near the source of the sound.

1 What is 'noise'?

2 How much louder is a 40 dB sound compared with a 30 dB sound?

3 Jamal likes to listen to music with earphones. What two things could he do to reduce the risk of damage to his hearing?

4 Suggest two jobs for which people should wear ear defenders.

5 One reason why truck drivers should not drive for long periods is because they may get tired and fall asleep. Suggest another reason.

- Sound levels are measured in decibels (dB).
- Loud sounds can damage your hearing over a period of time.
- You can reduce the risk of damage by making the sound quieter, shielding your ears, reducing the time you are exposed, or moving away from the sound.

Loudness, amplitude, and oscilloscopes

Describing waves

Imagine a string attached to a wall. You can make a wave in it by moving your hand up and down. Your friend takes a picture of the wave so that the wave is frozen in time. How can you describe the wave that you have made?

We can describe waves using three **properties** that all waves have: amplitude, wavelength, and frequency.

The distance from one point on a wave to the *same* point on the next wave is

↑ Some sounds, like fireworks are very loud

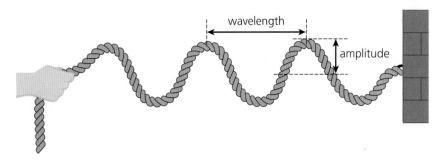

called the **wavelength**. This means that the distance from the *top* of one wave to the *top* of the next wave is one wavelength. So is the distance from the *bottom* of one wave to the *bottom* of the next wave.

The distance from the centre of the wave to the highest or lowest point is called the **amplitude**.

The diagram does not show the frequency. If you watched how many waves go past in 1 second, that would be the **frequency**.

Using an oscilloscope

Sound waves are made up of regions where the air molecules are close together and regions where they are far apart. A diagram like this shows the high-pressure regions (compressions) and the low-pressure regions (rarefactions).

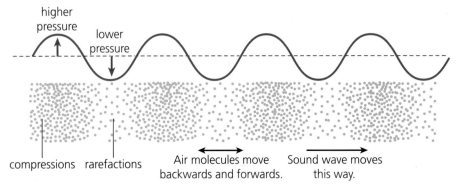

↑ A sound wave is pressure wave.

If we want to 'see' what sound waves are like we can use a microphone attached to an **oscilloscope**. The sound wave reaches the microphone which produces an electrical signal. This is displayed on the screen of the oscilloscope.

The picture on the screen shows how the pressure varies with time. Where the line goes up the particles are close together, and where it goes down they are far apart.

Loudness and amplitude

Anil sings a note into a microphone while he looks at the oscilloscope screen. Then he sings the same note, only louder. This is what he sees.

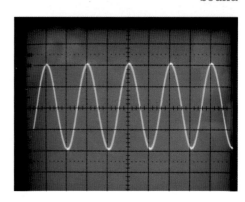

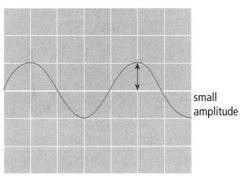

small amplitude

↑ A soft sound.

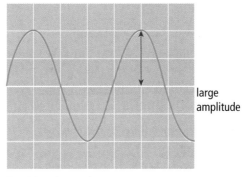

large amplitude

↑ A loud sound.

A soft sound has a small amplitude. A loud sound has a large amplitude.

Amplification

You might hear people say that sound 'dies away'. They think that the sound wave gets 'weaker' as it travels. The sound *spreads out* as it travels away from the source. Some energy is transferred to the medium, so the sound becomes softer.

A singer would not be able to sing loud enough for everyone in a big concert hall to hear her. She uses a microphone connected to an **amplifier** to make her voice louder. An amplifier increases the amplitude of the electrical signal produced by the sound wave, so that when it is broadcast through a loudspeaker it sounds louder.

1 Copy and complete this sentence.

A _____ sound has a larger amplitude than a _____ sound.

2 Sunil says that the amplitude of a wave is the distance from the centre to the top of a wave. Write a different definition of amplitude.

3 What does an amplifier change about a sound wave?

4 Which of these distances is 1 wavelength on a sound wave? There may be more than one correct answer.

 a the distance from a compression to a rarefaction

 b the distance from a compression to the next compression

 c the distance from a rarefaction to the next rarefaction

5 **Extension:** is a wave on the rope above a longitudinal wave or a transverse wave? How do you know?

6 The picture of the sound wave on the oscilloscope screen is a transverse wave. Sound waves are longitudinal waves. What is being shown on the screen?

- Waves have wavelength, amplitude, and frequency.
- An oscilloscope connected to a microphone can display a sound wave on a screen.
- The loudness of a sound depends on the amplitude.

5.5 Pitch and frequency

High and low

The rebab is a stringed instrument similar to a guitar that has been played for hundreds of years. It produces a range of notes that all have a different pitch, some high pitched and some low pitched. The **pitch** of a string depends on the number of times it vibrates every second.

The number of vibrations or waves per second is called the **frequency** and is measured in **hertz (Hz)**. High-pitched sounds have a high frequency and low-pitched sounds have a low frequency. High frequencies are measured in kilohertz (kHz).

Objectives

- Describe the link between pitch and frequency
- Know the range of hearing in humans
- Describe some of the differences between the range of hearing in humans and in animals
- Know why musical instruments are distinct

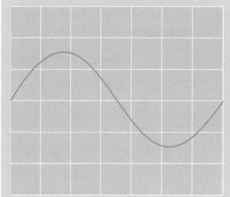

If the sound is low then the frequency is low.

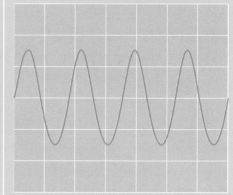

If the sound is high then the frequency is high.

The frequency is related to the wavelength. A high-frequency sound will have a short wavelength, and a low-frequency sound will have a long wavelength.

Changing the frequency (and wavelength) of a sound wave does not have the same effect as changing the amplitude. The amplitude determines how loud the sound is. A loud sound may be high pitched or low pitched.

What can you hear?

When you listen to speech or music you are listening to a range of frequencies, from very high to very low. Young people can hear a much larger range of frequencies than older people can. The range of human hearing is from 20 Hz to 20 000 Hz, known as the **audible** range. As people get older their audible range changes and they find it difficult to hear very high frequencies. By the time you are 30 you may be able to hear only up to about 14 000 Hz.

Ultrasound

Ultrasound is sound with a frequency that is higher than 20 000 Hz – outside the audible range. We cannot hear sounds of this frequency, or make sounds so high pitched, but other animals can.

What can other animals hear?

Species	Audible range (Hz)
bat	2000–110 000
cat	45–8000
chicken	125–2000
cow	23–35 000
dog	67–45 000
dolphin	100–100 000
elephant	16–12 000
guinea pig	54–50 000
horse	55–33 500
mouse	1000–91 000
rat	200–76 000
whale	1000–123 000

Different animals can hear very different sounds.

The sound of music

You may listen to music, or play an instrument yourself. When you listen to a song you can easily tell the difference between the voice of the singer and the instruments. If you heard exactly the same note being played on different instruments, you would be able to tell the instruments apart.

This is because as well as the main frequency of the note (the **fundamental**), there are other frequencies called **harmonics** mixed in with the note. It is the number and type of harmonics that make each instrument distinctive. The distinctive sound that an instrument makes is called the **timbre**.

Here is a note of the same frequency, 256 Hz, played on two different instruments:

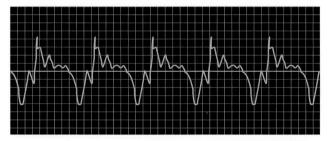

↑ A note played on a saxophone.

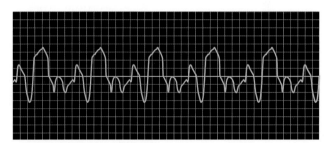

↑ A note played on a violin.

1 Aditya plays a note on a guitar. The note has a frequency of 512 Hz. How many times a second does the string vibrate?

2 Copy and complete the sentences.

   The pitch of a sound depends on the _____ of the sound.
   A high-pitched sound will have waves that have a _____ wavelength.

3 Sort these frequencies into those that are sound and those that are ultrasound.

   4500 Hz   100 000 Hz   30 000 Hz   30 Hz   1500 Hz

4 Which animal in the table above has the biggest range of hearing?

5 What is the name for the frequencies that make a musical note distinct?

- Higher notes have a higher frequency.
- The range of frequencies that you can hear is the audible range.
- Different animals have different audible ranges.
- Your audible range gets smaller as you get older.

Making simple calculations

Measuring the speed of sound

Diya wants to measure the speed of sound. She and Pari stand a distance of 165 m from the school building. Pari starts to clap, and hears the **echo** of her claps as they are reflected off the wall of the big building. An echo is a **reflection** of a sound wave from a surface. Pari adjusts her clapping until the clap coincides with the echo.

Diya takes a stopwatch to measure the time between each clap. To make a more **accurate** measurement she times how long it takes Pari to clap 10 times, and then divides that time by 10.

The time that Diya calculates is 1 second per clap.

Pari realises that the sound has travelled 330 m in 1 s, because it went all the way to the wall and back to her ears. She works out the speed of sound like this:

$$\text{Speed} = \frac{\text{distance}}{\text{time}}$$

$$= \frac{330 \text{ m}}{1.0 \text{ s}}$$

$$= 330 \text{ m/s}$$

Diya realises that the time it takes for the sound to reach the wall is actually 0.5 s. She works out the speed like this:

$$\text{Speed} = \frac{\text{distance}}{\text{time}}$$

$$= \frac{165 \text{ m}}{0.5 \text{ s}}$$

$$= 330 \text{ m/s}$$

You can work out the speed using either method. You need to remember that the time from a clap to its echo is the time that it takes the sound to reach the wall and come back again.

Thunder and lightning

Sound takes time to travel. During a thunderstorm we hear the thunder *after* we see the lightning, even though they are produced at the same time. This is because light travels much, much faster than sound.

Sound travels at about 330 m/s or 1225 km/h, so it takes about 0.3 s for sound to travel 100 m. A sprinter takes about 10 s to travel the same distance, so you might think that sound travels very fast.

The speed of light is much, much faster than the speed of sound. It would take light only about 0.000 000 3 s to travel 100 m. Light travels at 300 000 km/s or 300 000 000 m/s.

Maria is watching a thunderstorm. She learned in her science lesson that the thunder and lightning are produced at the same time, but that you hear the thunder *after* you see the lightning because it takes time for the sound to travel from the storm cloud to your ears. The time that it takes light to travel the same distance is so small that we say it arrives immediately.

She counts 9 seconds between the lightning and the thunder and works out the distance to the storm. The speed of sound is 330 m/s, which means that the sound travels 330 m each second.

Distance = speed × time
$$= 330 \text{ m/s} \times 9 \text{ s}$$
$$= 2970 \text{ m or } 2.97 \text{ km away}$$

Travelling faster than sound

Some vehicles can travel faster than the speed of sound. We say they are travelling at a **supersonic** speed. It is sometimes called 'breaking the sound barrier'. Scientists say that something travelling at the speed of sound is travelling at **Mach** 1. A plane travelling at Mach 2 has a speed of more than 660 m/s.

In October 1997 a car called *Thrust SSC* set a land speed record of 341 m/s or 1230 km/h. This was the first time that a car had travelled faster than the speed of sound. In October 2012, Felix Baumgartner became the first skydiver to travel faster than sound. He jumped out of a balloon 39 km above the Earth and reached a maximum speed of 1342 km/h, or Mach 1.1!

Q

1 How many times faster is the speed of light than the speed of sound?

2 Daniel is standing a certain distance from a wall and claps. He hears the echo 1.5 s later. How far away is the wall? Remember that the sound has travelled to the wall and back.

3 What will Maria notice when the storm is directly overhead?

4 In a 100 m sprint race, there are loudspeakers behind the starting blocks. Explain why.

5 **Extension:** which of these speeds are supersonic?

   200 m/s   300 km/h   400 m/s   1500 km/h

!

- An echo is a reflection of sound.
- When calculating the speed of sound using echoes, you use twice the distance or half the time.
- You can work out how far sound has travelled by multiplying the time and the speed.

Echoes

What is an echo?

If you stood in a very big cave and shouted, you would hear an echo of your voice. The sound spreads out and reflects off surfaces like the walls of the cave. Sound travels relatively slowly, so there is a time delay between your shout and the echo you hear when the sound is reflected from the walls of the cave.

Reducing echoes in buildings

Echoes can be a nuisance inside large rooms such as concert halls and theatres. The sound reflects off all the surfaces and echoes can last for several seconds, muddying the sound. This is called **reverberation**. If you are listening to a concert you don't want to keep hearing echoes, so the walls are designed to absorb a lot of the echoes. Theatre walls and ceilings are covered with soft, sound-absorbing materials.

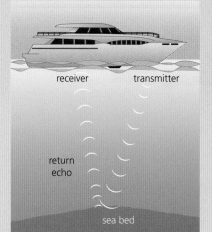

Sonar

In the past many ships used to be shipwrecked because they did not know whether there were rocks underneath the water. They used to lower a rope with a weight on the end, called a plumb line, to find the depth. Now ships are fitted with **sonar** that tells them what is underneath the ship. The word 'sonar' comes from **so**und **n**avigation **a**nd **r**anging. It can be used by submarines for navigating underwater.

The ship has an underwater loudspeaker (a **transmitter**) that produces pulses of ultrasound. This is sound waves with a frequency higher than 20 000 Hz. Beams of ultrasound are more focused than beams of audible sound, and they will not be confused with sounds made by people, animals, or other boats.

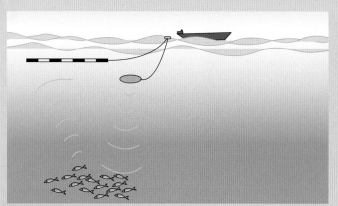

The ultrasound waves travel through the water and reflect off objects such as rocks or the sea bed. The echoes returning to the ship are detected by underwater microphones (the **receiver**). The sonar device uses the time of the echo being received to work out the depth of the water.

Sonar is very useful for fishermen. They can use the transmitter and receiver to find out where there are large shoals of fish. Finding the distance to an object in this way is call **echolocation**.

Animal echolation

Animals such as whales and dolphins use echolation more than they use sight. This enables them to find something to eat or find their family members even if it is dark or the water is not very clear. Sound travels much faster in water than in air, so they can send and receive sound messages over very long distances.

The dolphin makes a series of clicking sounds which reflect off a solid object such as a fish. The dolphin detects the echoes, and works out how far away the fish is using the time it took to detect the echo. Bats and other animals also find the distance to objects using echolocation.

Using ultrasound

We cannot hear ultrasound, but it is very useful. Pregnant women should not have X-rays because they could harm both the fetus and the mother. Instead doctors take ultrasound pictures to check the development of the fetus. The scanner uses ultrasound with a frequency of about 3 million Hz to make an image using the same principles as sonar.

The transducer transmits ultrasound into the woman, and it reflects off the fetus. The same transducer detects the echo and uses the time delay to work out where the ultrasound was reflected from. Ultrasound is reflected by the boundaries between soft tissue as well as from hard surfaces such as bone. By taking lots of pictures it is possible to build up a three-dimensional picture of the baby.

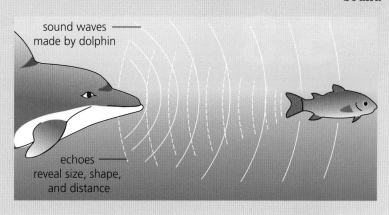

sound waves made by dolphin

echoes reveal size, shape, and distance

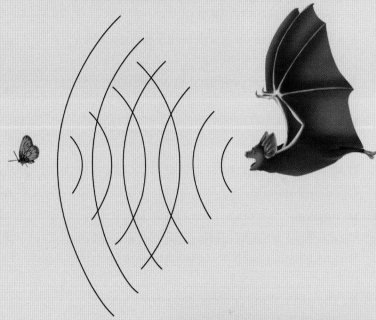

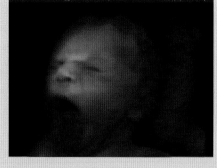

1 Describe how a fisherman can use sonar to find a shoal of fish.

2 Sound travels at 1500 m/s in water. How deep is an ocean if the echo from a transmitter takes 4 s to reach the receiver? (*Hint:* remember that 4 s is the time it takes to go to the sea bed and back.)

3 A dolphin makes a clicking sound and detects the echo from a fish that is 250 m away. How long was it before it heard the echo?

4 Why can dolphins find food at greater distances than bats?

5 A woman is having an ultrasound scan. Ultrasound travels at 1500 m/s inside the body. The ultrasound transmitter sends out a pulse of ultrasound and it is reflected from a surface 7.5 cm away. How long before the echo is received (Remember: 7.5 cm = 0.075 m)?

- We use echoes to find the distance to objects.
- Animals use echoes to locate and identify their prey.
- Ultrasound is very-high-frequency sound that can be used to make an image of a fetus.

1 Match each word with its correct definition. Write out the numbers and letters. [4]

Definitions

1 A region where the particles are close together in a sound wave is called a …

2 The solid, liquid, or gas that the sound wave is moving through is called the …

3 A region where the particles are far apart in a sound wave is called a …

4 A sound wave is produced by something that is ….

Words

A vibrating

B rarefaction

C compression

D medium

2 Think about what happens when you are talking. Decide whether each of these statements is true or false.

a The air molecules move away from your mouth as you talk. [1]

b The sound wave makes the molecules in the air move backwards and forwards as you talk. [1]

c The sound wave makes the molecules move up and down as you talk. [1]

3 The speed of sound in steel is 5000 m/s, in water is 1500 m/s, but in air is only 330 m/s.

Copy and complete the sentences below, using these words or phrases to fill in the gaps. Some words or phrases are not needed.

closer together   further apart   less   more

a The speed of sound in water is faster than the speed of sound in air because in water the particles are _____ than they are in air. [1]

b The speed of sound in air is slower than the speed of sound in steel because in steel the particles are _____ than they are in air. [1]

c A sound made underwater will take _____ time to travel a certain distance than a sound made in air. [1]

4 The table shows the speed of sound in each of six materials. Use the table and the information in question 3 to answer the questions below.

Material	Speed of sound (m/s)
A	4000
B	1207
C	1497
D	259
E	6420
F	435

a Which two materials are probably liquids? [1]

b Which two materials are probably gases? [1]

c Which two materials are probably solids? [1]

d In which material are the particles closest together? [1]

e In which material are the particles furthest apart? [1]

5 Look at the arrows on the diagram below. Copy the table and tick the correct columns to show whether each arrow shows the wavelength, the amplitude, or neither.

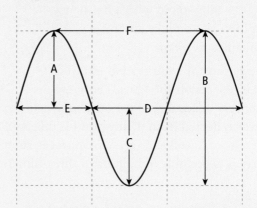

Arrow	Wavelength?	Amplitude?	Neither?
A			
B			
C			
D			
E			
F			

[6]

6 A boy is whistling a note that has a frequency of 1500 Hz.

a Explain what 'a frequency of 1500 Hz' means. [1]

b What would change about the sound he would hear if he whistled at 2000 Hz? [1]

7 Rani used a sound-level meter to survey the noise at different places and times during the day. Here are her results.

Place	Noise level (dB)
by the road on the way to school	70
sitting in a classroom	50
playing with friends	60
traffic on the way home	80
reading a book	40

a What is meant by 'dB'? [1]

b Which sound would sound the loudest? [1]

c Rani's brother likes to listen to very loud music on his headphones. Describe two things that he could he do to reduce the risk of damaging his hearing. [2]

8 A teacher uses an oscilloscope to display some sound waves. The sounds shown are either loud or soft, and either high or low pitched. Match the number to the letter in the pictures below.

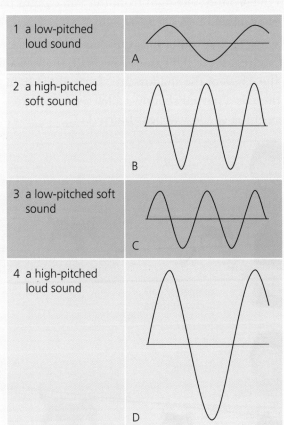

1 a low-pitched loud sound — A

2 a high-pitched soft sound — B

3 a low-pitched soft sound — C

4 a high-pitched loud sound — D

[4]

9 For each statement about ultrasound, write 'true' or 'false'.

a Ultrasound can be used to see an unborn baby. [1]

b Ultrasound is sound with a frequency greater than 2000 Hz. [1]

c Ultrasound is very-high-frequency sound. [1]

d Ultrasound cannot be heard by humans. [1]

10 A fisherman uses sonar to find a shoal of fish. A pulse of ultrasound is sent out and the reflection is detected 0.4 seconds later.

a How long did it take the sound to travel from the boat to the fish? [1]

b The speed of sound in water is 1500 m/s. How far below the boat are the fish? [1]

c Sometimes fishermen will see another signal on the sonar after they detect the echo from the fish. Why? [1]

11 Adami is watching a thunderstorm. She counts 4 seconds between seeing the lightning and hearing the thunder. The speed of sound in air is 330 m/s.

a How far away is the storm? [2]

b What assumption has Adami made to do this calculation? [1]

12 A teacher uses an oscilloscope to show the same note being played on a saxophone and a violin. Here is what is shown on the screen:

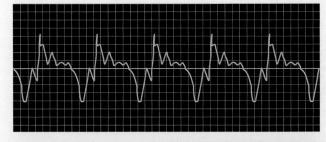

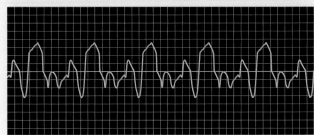

a How can you tell that they are playing the same note? [1]

b How can you tell that they are two different instruments? [1]

6.1 What is light?

Objectives

- Describe what light is
- Explain how shadows and eclipses form
- Describe how a camera works

Light transfers energy

Light is a way of transferring energy. **Light sources** give out light. We say they are **luminous**.

Energy is transferred from the Sun to the Earth by light waves. Light can reach us from the Sun and other stars. It can travel through empty space because unlike sound, light does not need a medium to travel through. **Infrared radiation**, called heat or **thermal energy**, is very similar to light and is another method by which energy reaches us from the Sun.

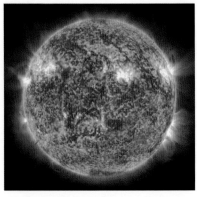

⬆ The Sun is a natural source of light.

Shadows

Shadows are dark areas where something blocks out light. They are formed because light travels in straight lines and does not bend round corners, so if something is in its path the light is blocked. On a sunny day you can look down and see your shadow on the ground. The light from the Sun cannot pass through you because you are **opaque** and block out the light.

Sometimes shadows have fuzzy edges and sometimes they have sharp edges. It depends on the type of light source. If the light source is small, there is a full shadow called an **umbra** and the edges of the shadow are sharp. If the light source is large or spread out, there is a partial shadow called a **penumbra** as well as an umbra and the edges of the shadow appear blurred.

⬆ A puppet theatre uses puppets to make shadows.

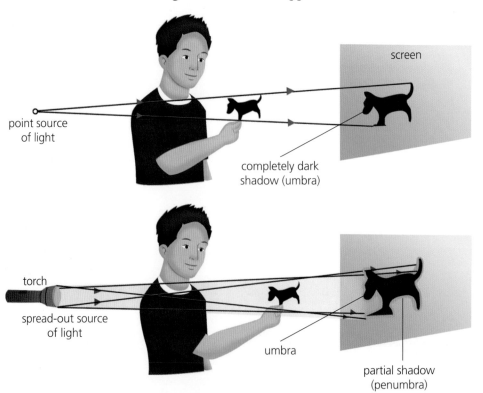

screen

point source of light

completely dark shadow (umbra)

torch

spread-out source of light

umbra

partial shadow (penumbra)

Line of sight

You can demonstrate that light travels in straight lines using holes cut in card. You cannot see the candle unless all three holes are exactly lined up.

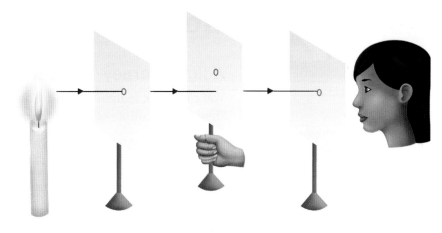

Cameras

If you take a picture of a friend with a camera, light is reflected from your friend and goes into the camera. Your friend is the **object**, and an **image** is formed at the back of the camera. The image is formed because light travels in straight lines.

A pinhole camera is a very simple camera. It is a box with a very small hole at the front made with a pin. At the back is a translucent screen where you can see the image. The image is not very bright because only a small amount of light can get through the hole.

Over 1000 years ago Al-Haytham, an Arab scientist, explained how a pinhole camera works. The light from the top of the **object** forms the bottom of the image. The image is upside down, or **inverted**. An image that is the right way up is **upright**.

In the past cameras used film to capture the image. Modern cameras use electronic devices called **CCDs** to produce an image that can be stored on a computer. You will find out more about CCDs on page 185.

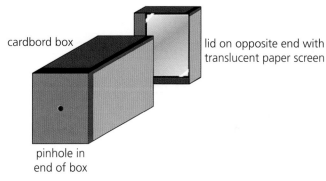

cardbord box

lid on opposite end with translucent paper screen

pinhole in end of box

⬆ A simple pinhole camera.

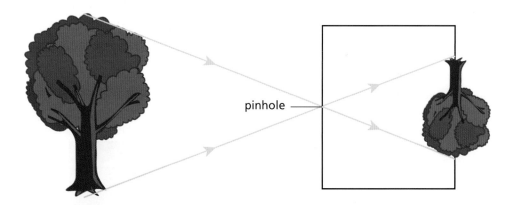

pinhole

Q

1 Name two luminous objects other than the Sun.

2 What can a shadow tell you about the size of the source of light?

3 Ammon is playing with a pinhole camera. He sees an image of a lamp on the screen of his camera.

　a He moves the camera towards the lamp. What two things would he notice about the image?

　b He holds the camera still and makes a very big hole in the front of the camera. The image is now very fuzzy. What else would be different about the image? Explain your answer.

- Light waves transfer energy from sources such as the Sun.

- Anything that gives out light is luminous. Shadows are formed because some objects are opaque.

- Light travels in straight lines. This explains how shadows and images are formed.

- A camera forms an inverted image on a screen.

How do we see things?

A light journey

You can think of light travelling on a journey from a **source** to a **detector**. Here is an example of a light journey.

- If light is given out by an object we say that the light is **emitted**.
- If light passes through an object we say that it is **transmitted**.
- If light is bounces off an object we say that it is **reflected**.
- If light is not transmitted or reflected but stays inside the object, we say that it is **absorbed**. (Absorbing light makes the object heat up a bit.)

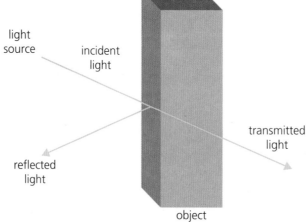

Materials behave differently when we try and see through them. These words describe some ways they behave with light.

You can see through clear glass. Glass is described as **transparent**.	Frosted glass lets light through, but you cannot see through it. Frosted glass is described as **translucent**.	Bricks do not let light through them. They are described as **opaque**.

Light level or **light intensity** can be measured using a lightmeter. Light intensity is measured in units called **lux**. It is a measure of the light energy hitting a certain area per second. The light level at midday is about 20 000 lux. In a bright room it is about 80 lux, and during a full moon it is about 1 lux.

The light spreads out as it moves away from the source. This means that the light hitting a certain area will be less, so the light intensity will be less. Some people think that light 'runs out' but that is not the case. The source appears dimmer as you move away from it because the light is spreading out.

Seeing things

The Sun is luminous. It gives out, or emits, light. In the same way, you can see other stars at night because the light from them enters your eyes.

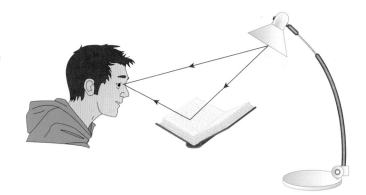

Objects that don't give out light are **non-luminous**. You can see them because light is reflected from them into your eyes. Sometimes people think that only mirrors reflect light, but all objects reflect at least some light. This is how we see them.

Black and white

A white shirt reflects most of the light shining on it. A dark shirt absorbs most of the light shining on it, and so our eyes do not detect very much light reflecting off it. The lack of light is what we 'see' as dark or black.

The eye

The **eye** is the human body's camera. If you look at a tree, the light from the Sun is reflected off the tree into your eye. The light enters your eye through the **pupil**. It is focused onto the back of your eye by the **cornea** (the transparent outer layer of the eye) and the **lens** (a small sack of jelly-like material).

The image of the tree is formed at the back of your eye on the **retina**, where light-sensitive cells called **rods** and **cones** absorb the light. Rods are mainly around the edge of the retina and are sensitive to dim light. Cones are mainly found in the centre of the retina and they are sensitive to bright light and colour. You can find out more about light-detecting cells in your *Complete Biology for Cambridge Secondary 1* Student book.

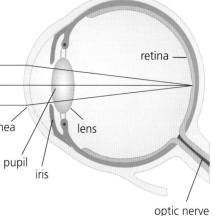

The rods and cones convert the light into a signal that is sent to the brain along the **optic nerve**. Like the image on a camera screen, the image on the retina is inverted. Your brain sorts it out so that you see things the right way up.

Q

1 Explain the difference between the words 'emit' and 'transmit'.

2 Is the Moon a luminous or a non-luminous object? Explain your answer.

3 Why is it harder to see dark-coloured cars at night than light-coloured ones?

4 **Extension:** the eye works in a similar way to the pinhole camera. Copy and complete the table below to compare the eye and the camera.

	Eye	Pinhole camera
hole to let the light in		
where the image is formed		

5 **Extension:** describe one way in which the eye is not like a camera.

!

- Sources emit light, and detectors absorb light.

- Light can be reflected, transmitted, and absorbed.

- Materials can be transparent, translucent, or opaque.

- When light is absorbed the object heats up.

- The eye works like a camera.

The speed of light

Objectives

- Know how fast light travels
- Understand how astronomers use the speed of light to describe distances

How fast does light travel?

The **speed of light** is the fastest speed that there is. Nothing travels faster than light.

There are lots of ways of writing down the speed of light.

300 million m/s   300 000 000 m/s   300 000 km/s   300 thousand km/s

Using speed to measure distance

Rajiv reads that light takes 1.3 seconds to travel from the Moon to the Earth. He wonders if he can work out how far away the Moon is.

$$\text{Speed} = \frac{\text{distance}}{\text{time}}$$

so distance = speed × time

= 300 000 000 m/s × 1.3 seconds

= 390 000 000 m or 390 000 km

In astronomy distances are enormous. For example, if you travelled to the Moon at the speed of an aeroplane, 900 km/h, it would take you nearly 18 days. To travel to the Sun would take 19 years, and to travel to Neptune would take nearly 600 years!

Here are those distances written in kilometres:

- Earth to the Moon: 384 400 km
- Earth to the Sun: 150 000 000 km
- Earth to Neptune: 4 500 000 000 km.

These large numbers can be difficult to work with. Instead of measuring distance in kilometres, astronomers use the distance that it takes light to travel in one year. This is called a **light year**. It seems strange to talk about a distance using units of years, but a light year is a measure of distance not time.

A light year is approximately 9 000 000 000 000 km. The distance to our nearest star, Proxima Centauri, is about 4.1 light years, and the diameter of the Milky Way galaxy is 100 000 light years. It is much more convenient to write down these huge distances in light years rather than metres or kilometres.

Here are some distances in light time.

Place	Distance in light time
Moon	1.3 light seconds
Sun	8 light minutes
Mars	between 4 and 20 light minutes
furthest that a spacecraft has travelled	18 light hours
nearest galaxy	2.5 million light years

Because light takes time to travel from faraway stars and planets, when astronomers study them they are not seeing them as they are now. They are seeing the Sun as it was 8 minutes ago. If you look up at the night sky, you are seeing stars and galaxies as they were when the light left them. For some objects this is thousands or millions of years ago!

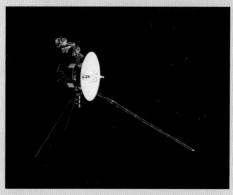

↑ *Voyager 1* has travelled further in space than any other spacecraft.

Measuring the speed of light

For thousands of years scientists argued about how long it takes light to travel. With sound you notice a time delay, but you cannot detect a delay between light being emitted and light being detected. Over 2000 years ago the Greek philosopher Empedocles thought that light did take time to travel, but another Greek philosopher Aristotle disagreed – he thought that light travels instantaneously.

In 1021, the Islamic philosopher Al-Haytham argued that, because light reflects off objects into your eye, it must take a certain amount of time to do so even if we cannot sense it.

↑ Andromeda is the nearest galaxy to the Milky Way.

It was not until there were methods for measuring very, very short periods of time that accurate measurements of the speed of light could be made.

The metre

For nearly 200 years there was a piece of metal that was kept in Paris, France that was exactly 1 m long. People would check the length of their rulers against this bar. In 1983 the definition changed. An international committee decided that one metre is 1/299792458 of the distance that light travels in 1 second. So the units of distance that we use are related to the speed of light.

Q

1 Why do astronomers use distances in light years rather than kilometres?

2 Look at the table of light times. It takes light about 9.3 times as long to reach us from Saturn compared with the Sun.

  a How far away is Saturn in light minutes?

  b Approximately how long would it take to get to Saturn travelling at the speed of an aeroplane?

3 Explain why looking up at the stars at night is like looking back in time.

4 Mars and Earth both orbit the Sun. Use this information to explain why there are *two* distances from the Earth to Mars in the table above.

5 What might be one of the problems with using a metal bar as the standard length for the metre?

- Light travels at 300 000 km/s. Nothing travels faster than light.

- It is convenient to measure the distance to very distant objects using light years rather than metres or kilometres.

- Measurements of the speed of light are difficult because the times are so short.

Reflection

Looking in a mirror

We often see reflections. Water can behave like a mirror and reflect animals or the sky.

Objectives

- Describe how an image in a plane mirror is formed
- Describe the differences between you and your image
- Explain why you see your image only in certain situations

Adinda is looking in a flat mirror, or **plane mirror**. We call her reflection an **image**. It is the right way up, the same size, and the same colour as Adinda. If she moves closer to the mirror the image moves closer, and if she moves away the image moves away. The image always appears to be the same distance behind the mirror as Adinda is in front of it.

There is nothing behind the mirror – no light can get through it. We say that the image is **virtual** – another Adinda does not exist behind the mirror. She needs to look in the mirror to see her image.

If you go to watch a movie, the pictures are projected onto a screen. Images that are projected onto a screen are called **real** images. The image in a pinhole camera is also real.

If Adinda holds up her left hand, the image she sees in the mirror appears to be holding up her right hand. We say that the image we see in a mirror is **laterally inverted**.

In these pictures the word AMBULANCE is laterally inverted on the front of the vehicle, but not on the back.

How is the image in a mirror formed?

Light from an object such as a candle reflects off the mirror into your eyes. Your brain knows that light travels in straight lines, so it assumes that the light has come from where it sees the image. A **ray diagram** shows how the image is formed. A ray diagram is a model of what is happening when the image is formed. Light spreads out from a source in all directions. Some of the light from the candle will be reflected from the mirror. There are millions of different paths that the light can take as it leaves the candle. We choose two paths that the light takes to help us to work out where the image is.

An image is formed when you look at yourself in the mirror. Light from a source like a light bulb reflects off your face, then off the mirror into your eye.

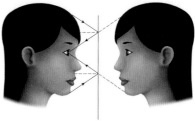

The light from the top of your face appears to come from the top of the image. The light from the bottom of your face appears to come from the bottom of the image. The image is the right way up.

Adinda's face   virtual image

Types of reflection

All surfaces, rough or smooth, reflect some light. The image that you see in a mirror is very clear. An image on water will only be clear if the water is flat. If the water's surface is very rough you do not see an image at all. The light must be reflected in a regular way to see an image.

A piece of white paper reflects most of the light that hits it, but you cannot see an image in it. This is because the surface is rough and the reflection is **diffuse**. The light is scattered from the surface.

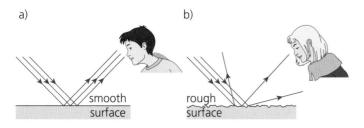

↑ Reflection from a smooth surface (regular reflection).

↑ Reflection from a rough surface (diffuse reflection).

Q

1 Which CAPITAL letters of the alphabet look the same in a mirror?

2 If you close your right eye, which eye does your mirror image close?

3 Explain why the word 'ambulance' is laterally inverted on the front of the vehicle but not on the back.

4 Different surfaces reflect light in different ways.

   a Explain why you can see your face in a shiny metal saucepan.

   b Sometimes you can see a faint reflection in the surface of a shiny plate or cup. Why?

   c Explain why you cannot see a reflection in a painted white wall.

!

- You see an image when light reflects from a smooth surface such as a mirror, water, or glass.

- The image is virtual and laterally inverted.

- You do not see images in rough surface because the reflection is diffuse.

117

Making measurements: the law of reflection

Objectives

- State the law of reflection
- Use the law of reflection
- Describe how to make accurate measurements in experiments with light rays

What is the law of reflection?

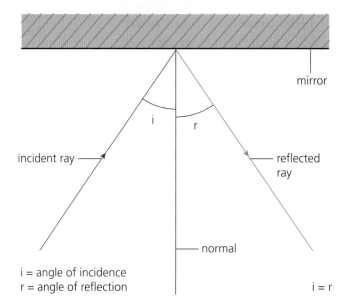

i = angle of incidence
r = angle of reflection

i = r

Light reflects off a smooth shiny surface, such as a mirror, in a regular way. A ray of light, called the **incident ray**, hits a mirror at a particular angle.

You can measure this angle. First you draw a line at right angles to the mirror at the point where the ray hits it. This line is called the **normal**. The **angle of incidence** is the angle between the normal and the incident ray, and is usually labelled *i*. (The angle of incidence is *not* the angle between the mirror and the ray.)

The ray reflects off the mirror. The angle between the normal and the **reflected ray** is the **angle of reflection**, labelled *r*.

The angle of incidence = the angle of reflection. This is the **law of reflection**.

Making measurements

Arjun and Diya are investigating the reflection of light in a mirror. They set up a ray box, a device that shines a ray of light. Arjun places a mirror to reflect the ray of light.

How are we going to take measurements?

We could draw the mirror and the normal on some paper.

Diya draws a diagram of the normal and the mirror. Arjun puts the mirror on top of the diagram and shines the ray box at the mirror. He makes sure the ray hits the mirror where the normal is drawn.

"How can we measure the angle of incidence and the angle of reflection?" wonders Diya. Arjun suggests drawing the rays on the diagram. He draws two dots on the paper under each ray.

He takes the ray box away and Diya joins up the dots to draw the rays. Now she can measure the angle of incidence and the angle of reflection.

They move the ray box to make different angles. Each time they draw the angle of incidence and the angle of reflection. Diya records their results in a table.

Angle of incidence (°)	Angle of reflection (°)
10	12
30	29
50	51
70	68

Mirrors in periscopes and kaleidoscopes

Periscopes can be useful things. A periscope enables you to look over the top of a wall, or over other people if you are at the back of a crowd.

Two mirrors at an angle will produce lots and lots of images. You can use this to make a **kaleidoscope**.

obstacle

⬆ Parallel mirrors are used to make a periscope.

Q

1 Name two things that Arjun and Diya did to get **accurate** measurements of the angles in their experiment.

2 Suggest one improvement that they could make to their method that would make the results more accurate.

3 Look at the drawing of the periscope. What are the angles of incidence and reflection at each mirror?

4 A student does an experiment with a mirror. She measures the angle between the *mirror* and an incident ray and finds that it is 30°.

   a What is the angle of incidence?

   b What is the angle of reflection?

!

- The angle of incidence equals the angle of reflection. This is the law of reflection.

- Mirrors are used in periscopes and kaleidoscopes.

- To take accurate measurements about the law of reflection you need to draw carefully where the rays go.

Refraction: air and water

Refraction of light

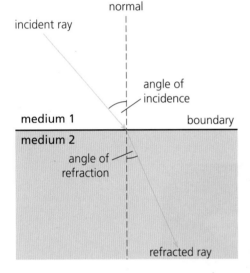

Is this pencil really bent? Why does it look bent?

When light travels from one substance to another, such as from air to water, it can change direction. This 'bending' of light is called **refraction**. The change of direction happens at the boundary, or interface, between medium 1 and medium 2.

The angle between the normal and the ray reaching the boundary is the angle of incidence. The angle between the ray and the normal on the other side of the boundary is the **angle of refraction**. If light goes from a less dense medium to a more dense medium, like from air to water, the direction of the ray will move *towards* the normal. If light goes from a more dense medium to a less dense medium, like from water to air, then it will move *away from* the normal.

Refraction explains why a pool looks shallower than it actually is. Light is reflected off a rock at the bottom of the pool and travels to the surface. It changes direction when it travels into the air.

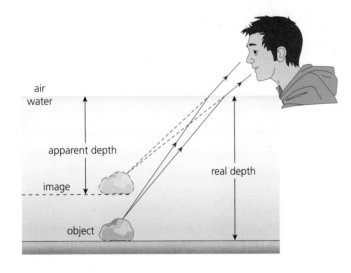

When light reaches your eye, your brain assumes that it has travelled in a straight line. So you see the rock as being in a different place. The depth that the rock appears to be is called the **apparent depth**, which is shallower than the **real depth**.

The bent pencil isn't really bent. The refraction of the light from the end of the pencil at A makes it look as if the light is coming from B.

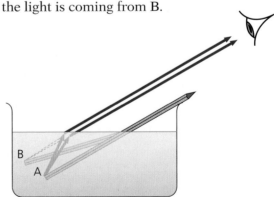

Why does refraction happen?

These people are marching side by side in a carnival. They are moving from tarmac onto sand. As each person reaches the sand they slow down. This changes the direction of the row, because there is a change in speed.

The same thing happens to light when it is refracted. In air, light travels at about 300 million km/s but in water it only travels at about 230 million km/s. So light going from air into water slows down. Light coming out of water into the air speeds up. In both cases the direction will change.

How much the speed changes in different materials depends on the **density** of the material. A more dense material will slow light down more, so the light will be refracted more.

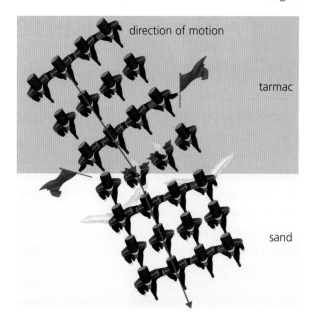

direction of motion

tarmac

sand

The speed of light and refraction

When light goes from air into a denser medium (material) such as water or glass, it slows down. The amount that it slows down depends on the material. Scientists work out a ratio for the two speeds called the **refractive index**.

$$\text{Refractive index} = \frac{\text{speed of light in a vacuum (empty space)}}{\text{speed of light in the medium}}$$

The speed of light in air is almost the same as the speed of light in a vacuum. Refractive index does not have a unit because it is the ratio of two speeds.

The refractive index is high if the light slows down a lot. The refractive index of diamond is 2.4, and the refractive index of glass is 1.5. Diamond slows light down a lot more than glass. This means light is refracted more when it goes into diamond than when it goes into glass.

Q

1 Think about the carnival marching into sand. If the rows marched straight towards the sand, and not at an angle, what would happen to:

   a their speed

   b their direction?

2 Diving birds often dive straight down if they want to catch a fish. Explain why.

3 A hunter is standing on the edge of a river trying to spear a fish. Should he aim above or below where he sees a fish? Explain your answer.

4 **Extension:** calculate the refractive index of water using the information above. What assumption are you making?

- Light changes direction when it goes from air into water, or water into air.
- This is called refraction.
- Refraction is a result of light slowing down.
- Pools appear shallower and objects appear bent in water because of refraction.

6.7 Refraction: air and glass

Objectives

- Use scientific knowledge to explain predictions
- Describe what happens when light goes through a glass block
- Explain total internal reflection

Investigating refraction

Daren is planning an investigation into refraction.

This is Daren's **prediction**.

This glass block feels a lot heavier than this plastic one. I wonder if there is a difference in how they refract light

> I predict that the glass block will refract the light more than the plastic block. This is because the glass is heavier than the plastic.

Daren's friend Amadi looks at Daren's prediction. Daren then rewrites it:

> I predict that the glass block will refract the light more than the plastic block.
>
> This is because the glass is denser than the plastic. When light goes from air into a dense material, it slows down and is refracted. Denser materials will slow light down more, so light will be refracted more by the glass than the plastic.

When you make a prediction you are not just saying *what* you think will happen. You should say *why* you think that will happen, using scientific knowledge. Daren can test his prediction by doing an experiment to collect data. He can use the scientific knowledge that he has to explain his **conclusion**. A conclusion says what you have found out.

You are right. I think that it is to do with the density. I have looked it up in a book and the glass is denser than the plastic.

But Daren, you haven't explained *why* you think that will happen using what we know about the speed of light.

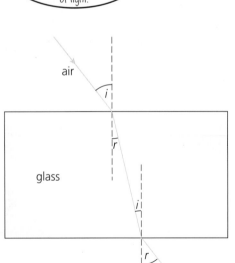

Refraction in a glass block

When a ray of light goes from air into glass, it is refracted.

- As the ray goes from the air into the glass it *slows down*. The ray bends *towards* the normal. The angle of refraction is *smaller* than the angle of incidence.
- As the ray goes from the glass into the air it *speeds up*. The ray bends *away from* the normal. The angle of refraction is *bigger* than the angle of incidence.

The ray going into the block is parallel with the ray coming out of the block. If the ray travels along the normal it does *not* change direction, but it will still slow down.

Total internal reflection

When light travels from a more dense substance into a less dense substance, such as from glass or water into air, the angle of refraction is *bigger* than the angle of incidence.

As the angle of incidence increases, so does the angle of refraction. Eventually the angle of refraction reaches 90°. The angle of incidence at this point is called the **critical angle**.

When the angle of incidence increases further to be larger than this critical angle, then all the light is reflected back into the glass or water. The refracted ray does not come out into the air again. This is called **total internal reflection**.

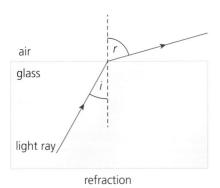

refraction

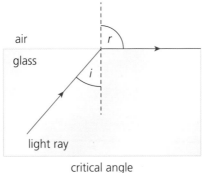

critical angle

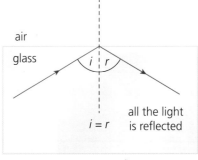

all the light
$i = r$ is reflected

total internal reflection

Using total internal reflection

An **optical fibre** is a very thin fibre made of glass or plastic. If you shine light into one end, the light is totally internally reflected down the fibre, which means that all of the light stays inside the fibre.

An **endoscope** is used to see inside the human body. Light is sent down one bundle of optical fibres, and shines on a structure inside the body. Another bundle of optical fibres transmits the light reflected from the inside of the body to a camera, so the doctor can see an image and diagnose the problem.

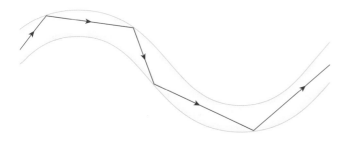

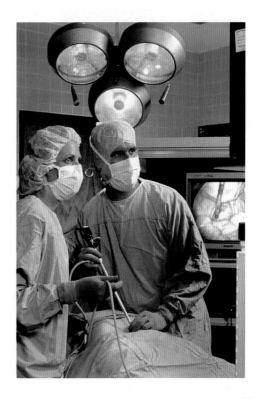

Q

1 Why was Daren's second prediction better than his first one?

2 Look at the diagram showing a ray going through a glass block. Explain why the ray leaving the glass block is parallel to the ray going into it.

3 Anyam drew a diagram to show how light travels down an optical fibre made of a material with a critical angle of 50°. What two things are wrong with his picture?

- You can justify predictions with scientific knowledge.

- Light is refracted when it goes from air into glass, or glass into air.

- If the angle of incidence is bigger than the critical angle, light is totally internally reflected.

Dispersion

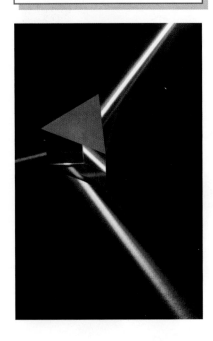

When you look at a normal light bulb it looks white. We use rays from a light bulb to investigate how light is reflected and refracted. This 'white' light is really a mixture of colours. We see those colours when we look at rainbows.

Splitting light

When light goes into glass or water it slows down. If a ray of white light shines into a rectangular block of glass it will change direction or **refract**. We see a ray of white light entering and leaving the block.

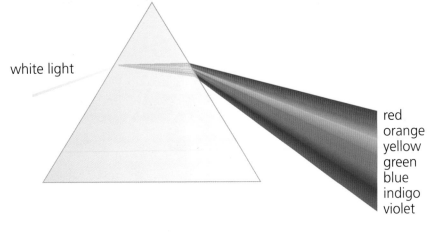

white light

red
orange
yellow
green
blue
indigo
violet

If the ray of light is shone into a triangular piece of glass, called a **prism**, again it slows down, but something different happens. When the incident ray hits the side of the prism at a particular angle, the light emerging from the prism is split into colours, called a **spectrum**. The process of splitting white light into separate colours is called **dispersion**. We cannot see the light inside the prism. From the colours that emerge we can work out the paths that the different colours take inside the glass.

The colours of the white light spectrum are:

red   **orange**   yellow   **green**   **blue**   **indigo**   **violet**

Dispersion happens because different colours of light are refracted by different amounts. Red light is refracted the least, yellow light is refracted more than red, and so on. Violet light is refracted the most.

How rainbows form

You see a **rainbow** when you are facing falling rain with the Sun behind you.

Raindrops behave like small prisms. The sunlight from behind you is refracted, reflected, and dispersed so that you see a spectrum, which we call a rainbow.

About 700 years ago people didn't know what caused a rainbow to form. An Iranian scientist, Al-Farisi, correctly explained why the raindrops dispersed the white light. One of his students demonstrated his idea by directing white light through a large spherical glass container filled with water.

Recombining the spectrum

More than 350 years later, in 1666, Sir Isaac Newton demonstrated that light from the Sun could be split into colours. He also showed that he could recombine all the colours back into white light. He did this by using two prisms.

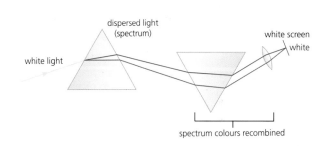

Explaining dispersion

A wave can be described by its frequency or wavelength. Light behaves like a wave. The different colours of light all have different wavelengths and different frequencies. Each colour is refracted by a different amount depending on its frequency. Higher frequencies such as violet are refracted more than lower frequencies such as red.

As the colours leave the prism they have all been refracted by different amounts. This spreads the colours out to produce a continuous spectrum.

Q

1 How many times is light refracted when it passes through a prism?

2 The spectrum of white light is described as continuous. What does that mean?

3 You can have refraction without dispersion, but you cannot have dispersion without refraction. Explain why.

4 Which colour of light travels fastest in glass? Explain your answer.

!

- A prism splits white light into the colours of the spectrum. This is called dispersion.

- A spectrum is formed when different colours are refracted by different amounts.

- The colours of the spectrum are red, orange, yellow, green, blue, indigo, and violet.

- Rainbows are formed when raindrops split sunlight into colours.

Colour

Colour addition – primary and secondary colours

In 1937 Ardeshir Irani made the first colour film in Hindi. It was not until the 1950s that colour film became popular. Now we are used to seeing films and television programmes in colour.

You can make all the colours of light with just three **primary colours**. The primary colours are **red**, **green**, and **blue**. If you combine any two of the three colours you get a **secondary colour**. If you mix all three primary colours of light you get white light.

- Red and green light make yellow light.
- Green and blue light make cyan light.
- Red and blue light make magenta light.

It is not possible to make red, green, or blue light using any combination of other colours. That is why they are called primary colours.

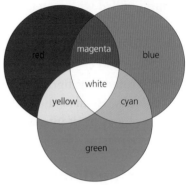

Colour displays on televisions and computers combine primary colours of light to produce the range of colours that you see.

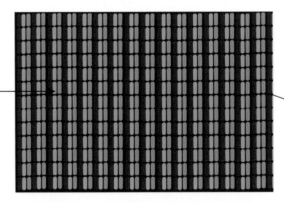

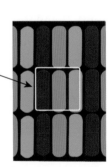

Colour subtraction – filters

Coloured lights in a theatre can make a production very dramatic. But if a light bulb produces white light, how can we produce coloured lights for the stage?

Coloured filters are used to produce different colours from white light. A **filter** absorbs some colours of light and transmits others. If you shine white light through a red filter, the filter will only let red light through. Red light is **transmitted** and all the other colours are **absorbed**. The filter has taken away all the other colours except red.

This means that the light you see through a filter will be dimmer than the light without the filter because some light has been absorbed.

If you shine red light through a green filter, then no light will get through. If you look through the red and green filter together you will see black. Your brain perceives no light as black.

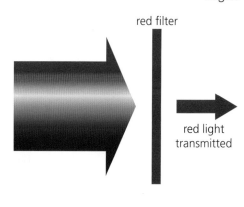

red filter

red light transmitted

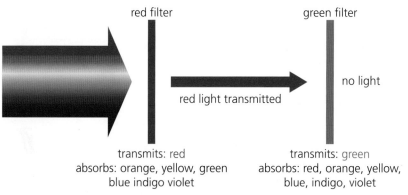

red filter

red light transmitted

green filter

no light

transmits: red
absorbs: orange, yellow, green
blue indigo violet

transmits: green
absorbs: red, orange, yellow,
blue, indigo, violet

⬆ If you combine filters of two primary colours, then no light will get through.

Filters do not just come in primary colours – you can also have a secondary colour filter. A magenta filter will transmit red and blue light and absorb green light. By choosing the right material for your filter you can produce a filter that will transmit a particular colour (wavelength) of light.

Your eye and colour

Your eye can detect different colours of light because of light-sensitive cells in the retina called cones. The cones absorb the light, and send a signal to your brain. Your brain processes the signals and produces the image that you see. The human eye can distinguish between thousands or millions of different colours.

Some people have cones that do not work properly and they don't see colours accurately. About 1 in 20 people (usually men) have a form of **colour blindness**. They find it hard to tell the difference between red and green.

If you have normal colour vision, you will see the numbers 7 and 4 in this picture. If you are red–green colour blind then you might see the numbers 2 and 1.

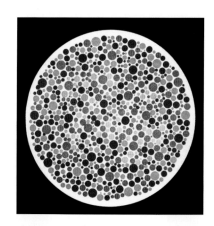

Q

1 Explain what a green filter does to white light.

2 Copy and complete the table to show what colour light would be produced in these situations.

Combining...	Makes...
red + blue	
cyan + red	
blue + yellow	

3 Some people think that coloured filters add colour to white light. Why is this not correct?

4 What happens if you pass white light through a blue then a yellow filter? Explain your answer.

!

● The primary colours of light are red, green, and blue.

● The secondary colours of light are yellow, cyan, and magenta.

● The primary colours add to give white light.

● Colour filters absorb all the colours of light except the colour that they transmit.

Presenting conclusions: more on colour

Objectives

- Explain why coloured objects look coloured in white light
- Explain why coloured objects look different colours in different colours of light
- Describe how to present conclusions in appropriate ways

More colour subtraction – coloured objects

⬆ The red shirt reflects red light and the blue shorts reflect blue light.

Markets sell fabrics of many different colours. Imagine that you are looking at a fabric that looks green. If the light from the Sun is a mixture of colours, why does only green light enter your eyes? Coloured objects *absorb* some colours of light and *reflect* other colours.

The red shirt is absorbing all of the colours in white light except red. The red light is reflected into our eyes and we see a red shirt.

An object looks white because *all* of the colours of light are reflected. An object looks black because *none* of the colours are reflected.

In science, the primary colours of light are red, green, and blue. The colours of light are added together to give all the other colours.

In art, the primary colours of paint are red, yellow, and blue. The different colours of paint absorb different colours of light. Green paint absorbs all the colours except green, which it reflects. It acts like a filter and subtracts the other colours.

If you mix all the colours of light together you get white light, but if you mix all the colours of paint together the paint looks black.

Coloured objects in coloured light

Filza and Aru are using filters to produce light beams of different colours. Aru shines coloured light on pieces of fabric of different colours. Here are her data.

	Red fabric	Blue fabric	Green fabric
Red light	red	black	black
Blue light	black	blue	black
Green light	black	black	green

Filza does some research on different coloured objects in different coloured lights. Here are her data.

	Red object	Blue object	Green object
Red light	red	black	black
Blue light	black	blue	black
Green light	black	black	green
Yellow light	red	black	green
Cyan light	black	blue	green
Magenta light	red	blue	black

Look at my results. My conclusion is: a red fabric will only look red in white light or red light. Otherwise it will look black. It's the same for green and blue.

I don't think that we have enough data to say that. There are lots of other colours of fabric and colours of light. Let's find some more data.

Some of the new data do support my conclusion, but some don't. Is there a better conclusion?

Filza writes this conclusion down.

> Coloured objects reflect the colours that they are. If the light shining on the object contains one or more of those colours, then that is the colour that you see. If the light shining on the object does not contain any of those colours, then it will look black. This is the case for both primary and secondary colours.

When you do an investigation and write your conclusion, you need to make sure that it takes into account *all* of the data. You should also make sure that you write a conclusion that your data support.

Q

1 You are looking at a yellow flower growing outside in the sunshine. Why does it look yellow?

2 Why does a combination of red, yellow, and blue paint look black?

3 A football team wears a yellow shirt and white shorts. What would the kit look like:

   a in red light?

   b in green light?

   c in blue light?

4 Which of the data in the second table do *not* support Aru's conclusion?

!

- A coloured object absorbs all the colours of light except the colour that it appears to be.

- Coloured objects appear different colours in different coloured light because they absorb certain colours and reflect others.

- When you write a conclusion, you need to make sure that your data support it.

Asking scientific questions

Objectives

- Understand that there can be different explanations for the same observations
- Explain why some explanations are accepted and others are not
- Understand that explanations change as new observations are made

Two ideas about light

Scientists make **observations** of the world around them. They describe what they see and try to make sense of their observations by asking questions and producing an **explanation**.

Sometimes two people who make the same observations can come up with different explanations. This was the case for two scientists who lived over 400 years ago: Isaac Newton and Christian Huygens.

⬆ Isaac Newton did lots of experiments with light.

⬆ Christian Huygens's idea about light was different from Newton's.

Waves and particles

Isaac Newton showed that light could be split into colours using a prism, and that it can be recombined with a second prism. He also invented a telescope that used mirrors instead of lenses.

Newton thought that light was made up of **particles**. Lots of people had used the same idea over thousands of years. Philosophers in India nearly 2000 years ago thought that light was made of a stream of high-speed 'atoms', and 1000 years ago the Arab scientist Al-Haytham wrote a book about light that included the idea that light was made of particles. Newton worked out explanations for lots of the observations about light using the idea of particles.

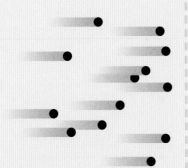

Christian Huygens thought that light was a **wave**. The idea that light is a wave is much newer than the idea of light as particles. Huygens thought that light waves moved in some invisible medium, just as waves move through water. He also worked out some explanations for observations about light.

Observation about light	Newton's explanation	Huygens's explanation
Light travels in straight lines.	Particles travel in straight lines.	Waves travel in straight lines.
Light is reflected at equal angles.	Particles of light bounce off the mirror like a ball off a wall.	Wave bounce off barriers in the way that sounds make echoes.
Light is refracted.	Particles near a boundary with a dense medium are pulled off course by the particles in the medium.	Waves slow down when they go into shallow water, and bend.

At the time the idea that light was a particle was very easy to understand because people could see balls bouncing off walls and imagine that light does the same thing.

But there was one observation that Newton found it very difficult to explain with his particle model.

When light passes into a glass block, some of it is refracted but some of it is reflected. It is difficult to explain this observation using the idea of particles of light.

Which idea was accepted?

Even though the particle idea did not explain some observations very well, people believed Newton's explanation for over 100 years. This was because Newton had published a lot of explanations about gravity and was a very famous scientist. All his explanations had been very successful at explaining observations that astronomers made about stars and planets.

In 1800 another scientist did an experiment on light that could not be explained with the idea of particles. The wave theory became accepted as a better explanation of how light behaves.

That is not the end of the story. Another 100 years later scientists did a light experiment and made some observations that could not be explained with the idea of waves! Now it seems that you need both ideas to describe what light is.

1 Why was the idea of particles popular?

2 Why was Newton famous?

3 You can use a football to model Newton's idea about light.

   a How could you use a football to show his explanation of the reflection of light?

   b Can you use a football to show how light is refracted? Explain your answer.

4 Newton tried to explain the fact that when light goes into a block of glass, some of it is reflected and some of it is refracted. He said that some particles of light bounce off and others go through the block. Is this a good explanation? Explain your answer.

- Scientists produce explanations for their observations about the world around them.
- Good explanations explain a wide range of observations.
- Sometimes explanations need to be changed if they do not explain what is observed.

Lasers

What is a laser?

Scientists often use light in their experiments.

A **laser** produces a narrow beam of light. The light that it produces is very different from sunlight.

Lasers produce light that is only one colour. We say that it is **monochromatic**. The light from a laser can be powerful enough to melt metal, but low-powered lasers are also very useful.

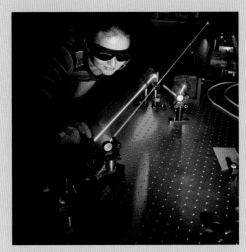

The first laser was built in 1960 by Theodore Maiman, an American physicist. Laser stands for:

Light

Amplification by the

Stimulated

Emission of

Radiation.

Maiman's laser used a ruby crystal. Later in 1960 an Iranian physicist, Ali Javan, and two others made the first gas laser using the gases helium and neon. Many lasers that are used in schools are helium–neon lasers.

Unlike ordinary light, the light waves in a laser are all in step. This means that it is possible to make a laser beam that is very powerful.

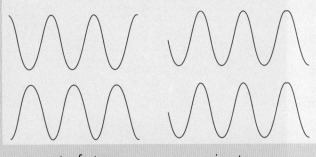

out of step   in step

Using lasers in industry

It is possible to make a laser that is so powerful that it can drill holes in metal. Laser light does not spread out, so it is very useful for welding metal with great precision. Even more powerful lasers can cut through steel. In clothing factories lasers can cut through many layers of fabric at the same time.

Using lasers for storing and reading data

Compact discs (CDs), DVDs, and Blu-ray discs are three types of data storage device that can store music, films, and other data. Blu-ray discs can store a lot of data – they can hold high-definition video and audio as well as photographs and other information.

All of the discs work in the same way. A laser makes a series of pits in the surface of the disc that codes for the music, film, or other data. The surface is then coated with a thin metal layer.

When you put the disc into a machine, a low-power laser bounces light off the surface of the disc and 'reads' the sequence of pits to reproduce the data. A computer decodes the data to produce the music or the film.

A barcode reader uses a low-power laser to identify products. The reader sends out laser light, and this is reflected off the black and white lines of the barcode on the product. The reflected laser light is decoded to give the name of the product.

Using lasers in medicine

One of the most common uses of lasers is to correct eyesight. Lasers can be used to change the shape of the cornea of your eye so that the light is focused on the retina. If someone's retina has become detached from the back of their eye then a laser can be used to 'weld' it back on again.

In the same way that lasers can cut metal, they can also be used to cut the tissues in the human body. A surgeon can use a laser scalpel. The laser beam is more precise than a traditional scalpel, and can be adjusted to cut at any angle.

1 How is laser light different from sunlight?

2 Describe one advantage, other than being able to cut many layers at the same time, of using a laser to cut material to make clothes.

3 A laser makes the pits in the surface of a DVD, and another laser is needed to read it. One is a low-power laser and the other is a high-power laser. Which is which?

4 Astronomers have used a laser to measure the distance to the Moon. Astronauts left a mirror on the Moon when they visited it during the *Apollo* missions.

  a Why do they need to use a laser rather than a regular light source?

  b Describe how they could use the laser to measure the distance to the Moon. (Hint: Think how you can use sonar to find the depth of the sea.)

- The waves in laser light are all in step.
- Lasers can be made that are powerful enough to cut skin or metal.
- Low-power lasers are useful in data-storage devices and to read barcodes.

1 Choose words from this list to match each definition below.

non-luminous inverted opaque penumbra umbra luminous upright translucent

a gives out light   [1]

b does not allow light to travel through it   [1]

c upside down (describing an image of an object)   [1]

d the very dark part of a shadow   [1]

e does not give out light   [1]

2 Copy and complete the sentences using the words from the list. You may need to use some words more than once.

reflected transmitted absorbed transparent translucent opaque refracted

a When light is _____ from a mirror you see an image.   [1]

b Light is _____ by objects that are opaque.   [1]

c If you hold up a piece of white cloth you can see light coming through it, but you cannot see through it. It is described as _____, not _____.   [2]

d You see non-luminous objects because light is _____ off them into your eyes.   [1]

e Wood and concrete are examples of materials that are _____.   [1]

f Water and glass are examples of materials that are _____.   [1]

3 Sanjay is holding his hand a distance from a lamp to make a shadow on a wall.   [1]

Which of these statements is *not* correct?

a If Sanjay moves his hand closer to the lamp, the shadow will get bigger.

b If he uses a dimmer lamp, the shadow will gets bigger.

c If Sanjay moves his hand closer to the wall, the shadow will get smaller.   [1]

4 Here are some diagrams showing reflection. Which pictures correctly show how light is reflected from a mirror?   [1]

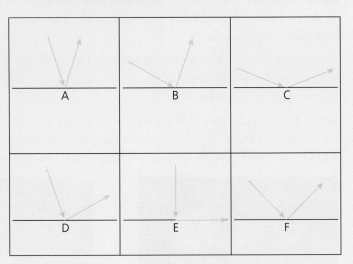

5 In which of these sentences are the words *emit, transmit, reflect, absorb* used correctly?

a A light bulb transmits light.

b A fire emits infrared radiation.

c A glass of water transmits light.

d A mirror emits light.

e You reflect light.   [1]

6 Look at the diagram of a ray of light being reflected.

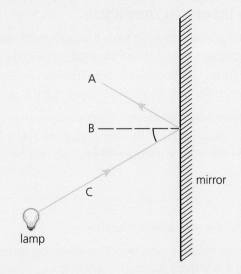

a Which line, A, B, or C, is the incident ray?   [1]

b Which line, A, B, or C, is the reflected ray?   [1]

c Which line, A, B, or C, is the normal?   [1]

7 This diagram shows what a fish sees from below the water.

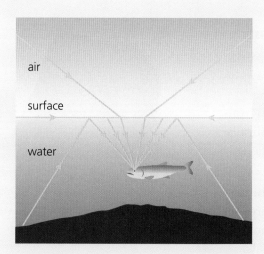

a What is the scientific name for light changing direction as it goes from air into water? [1]

b Why does the light change direction? [2]

c How is it possible for the fish see the bottom of the lake as well as the water's surface and the bank above the water? [2]

8 Sometimes prisms are used instead of mirrors to reflect light, for example in binoculars.

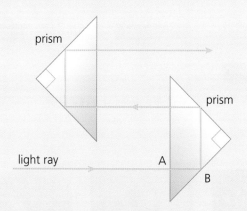

a What is the angle of incidence at A? [1]

b Explain what happens to the ray at A. [1]

c What is the angle of incidence at B? [1]

d Explain what happens to the ray at B. [1]

e What can you say about the value of the critical angle for the material that the prism is made from? [1]

f What would happen if this was not the case? [1]

9 Here are some statements about measuring the angle of refraction in a glass block.

A Put the glass block on a piece of white paper and draw around the glass block.

B Join the two rays with a straight line to show the ray inside the block.

C Remove the glass block, mark a point on one edge of the block, and draw a line at 90° to the edge at that point (the normal).

D Mark dots on the paper along the line where the ray goes on both sides of the block.

E Replace the block on the piece of paper and shine a ray of light at the block at the point where the normal meets the edge of the block.

F Remove the block and use a ruler to draw lines for both of the rays outside the block.

G Use a protractor to measure the angle of incidence and angle of refraction.

a Write the letters in order to show the correct order for the statements. The first and the last statements are in the correct place. [1]

b How does this method ensure that the measurements of the angles are as accurate as possible? [1]

10 Which of these statements about dispersion is correct? (There may be more than one correct answer.) [1]

A Dispersion happens when glass changes the colour of light.

B When white light is dispersed, it is split into different colours.

C Dispersion happens when white light is refracted in a prism.

D Dispersion only happens in prisms.

11 Jamila is in the front row watching an actor on a stage. She notices that the stage lights change the appearance of the actor's clothes. This is what she sees:

● In red light his trousers looks black and his shirt looks red.

● In green light his trousers look green and his shirt looks black.

● In blue light his trousers looks black and his shirt looks blue.

a What colour are his trousers? [1]

b What colour is his shirt? [1]

The properties of magnets

Discovering magnetism

A long time ago the Ancient Greeks discovered that a rock that they called **lodestone** attracted small pieces of iron. Lodestone is an iron-containing mineral that is naturally **magnetic**. The Ancient Greeks had discovered **magnetism**.

A magnetic force is a **non-contact force** – lodestone exerted a force on the pieces of iron even when they were not touching. Scientists can now make very strong **magnets** that float or levitate on some other materials.

Attraction and repulsion

Every magnet has a **north pole** and a **south pole**. The poles of magnets attract or repel each other.

When do two magnets repel?	When do two magnets attract?
• The north pole of a magnet **repels** the north pole of another magnet. • The south pole of a magnet **repels** the south pole of another magnet.	• The north pole of a magnet **attracts** the south pole of another magnet.
S N ← → N S N S ← → S N	S N → ← S N
• *Like poles repel.*	• *Unlike poles attract.*

Magnetic materials

⬆ Iron, steel, cobalt, and nickel are magnetic.

Some materials are attracted to magnets. These are called **magnetic materials**. **Iron**, **cobalt**, and **nickel** are all magnetic materials that are **elements**. Some of the **alloys** or **oxides** of these elements are also magnetic. You will learn more about alloys in your *Complete Chemistry for Cambridge Secondary 1* Student book.

Steel is magnetic because it contains mainly iron. Lodestone, or magnetite, contains iron oxide which is also magnetic.

The strip on credit cards is made of a plastic that contains lots of very small iron pieces that behave like tiny magnets. These iron pieces are used to store information on the card.

A piece of magnetic material is attracted to a magnet, but it will not be repelled by it. Magnetic materials do not attract or repel each other. The only two things that can repel each other are two magnets, so the way to tell whether an object is a magnet is to see if it is repelled by another magnet.

Why are some materials magnetic?

You can think of a magnetic material as being made up of lots of very small regions called **domains**. Each domain behaves like a tiny bar magnet.

When a magnetic material is not **magnetised**, the domains are pointing in lots of different directions. If you magnetise the material by putting it near a magnet or stroking it with a magnet, the domains line up. The material has been magnetised.

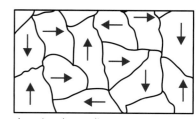

domains line up

domains do not line up

Magnetising and demagnetising

Iron and steel are magnetised by being close to a magnet. Iron is easy to magnetise, you can pick up iron nails with a magnet, but if you remove the magnet and hold them up, they are no longer magnetised so they fall.

Steel is harder to magnetise, but it stays magnetised when you remove the magnet. Steel pins or paperclips stay magnetised.

You can make a stronger magnet from a piece of magnetic material, such as a steel needle, by stroking the needle with a magnet repeatedly in the same direction.

A piece of iron or steel that has been magnetised is a **permanent magnet**. You cannot turn it off. You can **demagnetise** a magnet by heating it up or hitting it with a hammer.

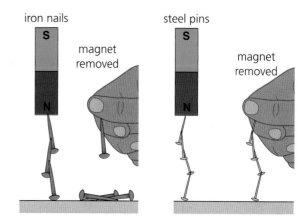

iron nails

magnet removed

steel pins

magnet removed

needle

Q

1 Sort the list below into materials that are magnetic and materials that are not magnetic.

brass   steel   iron   copper   wool   wood   nickel   cotton

2 You pick up some nails with your fingers from a pile of nails and find that they appear to stick together. Are they made out of iron or steel? Explain your answer.

3 Look at the picture. The two magnets are attracted to each other.

| A | B |→ ←| C | D |

a If A is a south pole, what poles are B, C, and D?

b What will happen in these two experiments?

experiment 1:

| A | B | | D | C |

experiment 2:

| B | A | | C | D |

4 **Extension:** Use the idea of domains to explain:

a why if you break a magnet in two you get two magnets each with a north and south pole.

b why you can demagnetise a permanent magnet by heating it.

c why iron does not stay magnetised when you remove the magnet, but steel does.

!

- Iron, cobalt, steel, and nickel are all magnetic materials that are attracted to magnets.

- For magnets, like poles repel and unlike poles attract.

- You can make a magnet by stroking a magnetic material with a magnet.

Magnetic fields

Ferrofluid is a special magnetic fluid that makes beautiful patterns when it is near a magnet. But why does it move?

What is a magnetic field?

If you put a steel pin near a magnet, the pin will move. The magnet attracts the pin because it is exerting a force on it.

The region around the magnet where magnetic materials experience a force is called the **magnetic field**. The ferrofluid moves because it is in a magnetic field.

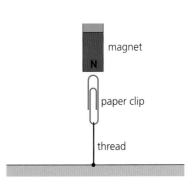

⬆ The paper clip is in the magnetic field of the magnet.

'Seeing' magnetic field patterns with iron filings

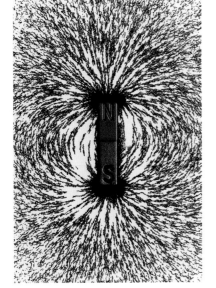

How can we 'see' a magnetic field? We can put something in the magnetic field that experiences the force to show us the shape of the field.

Iron filings are very small pieces of iron that become magnetised when they are in a magnetic field. They line up, just like nails or paper clips, and there are so many of them that they can show us the shape of the magnetic field. We cannot see the magnetic field, only the effect of the field on magnets or magnetic materials.

You can investigate the magnetic field patterns around two magnets.

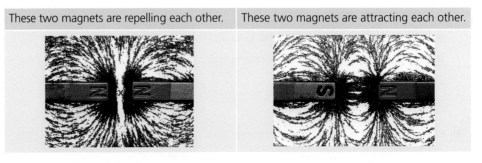

These two magnets are repelling each other.	These two magnets are attracting each other.

Point **X** is called the **neutral point**. At this point the magnetic fields of the two magnets cancel out.

'Seeing' magnetic field patterns with a plotting compass

If you suspend a magnet on a piece of cotton, the magnet will line up in the same direction each time. This direction is north–south. A **compass** points north because its needle is a small magnet. It rotates and lines up with the Earth's magnetic field.

A plotting compass is a very small compass. If you put it near a bar magnet, its needle will line up with the magnetic field of the magnet, rather than the Earth's magnetic field. You can use a plotting compass to see the shape of the field around a bar magnet.

We can show the pattern around a bar magnet more simply by drawing **magnetic field lines** to show where the plotting compass lined up. The lines show us the general shape of the field.

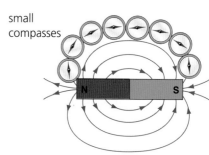

small compasses

- The spacing of the lines indicates the strength of the field. The field is stronger where the lines are closer together.

- The arrows on the lines show the direction of the force. Magnetic field lines go from the north pole to the south pole. The arrows could have pointed towards the north pole but scientists decided that they will always point towards the south pole.

Magnetic field lines are just a representation of the field. There is a field all around the magnet, and a magnetic material placed between the lines would still experience a force.

Although we only usually draw the magnetic field on a flat piece of paper, the field exists above and below the magnet as well.

The Earth's magnetic field

The Earth behaves as if there is a huge bar magnet inside it. The magnetic field pattern is like that of a bar magnet. But there is no bar magnet inside the Earth, and we are not exactly sure how its magnetic field is produced.

If a compass needle points north, then the *south* pole of the Earth's magnet must be in the north!

Compass needles actually point to the Earth's magnetic North Pole. This is not the same place as the Earth's geographic North Pole, which is at the top of the Earth's axis. The magnetic North Pole is in Canada, about 1200 km away from the geographic North Pole.

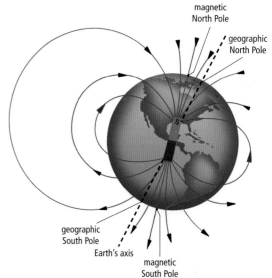

magnetic North Pole

geographic North Pole

geographic South Pole

Earth's axis

magnetic South Pole

1 What is a magnetic field?

2 Where is the magnetic field of a bar magnet strongest? How do you know?

3 Look at the pictures of iron filings around the two magnets repelling and two magnets attracting.

   a Draw simplified diagrams of the magnetic field pattern in each picture.

   b What would happen if you put a small steel ball at the neutral point?

   c Explain your answer.

4 A student says that he has made a magnetic field with some iron filings. What would you say to him?

- A region where magnetic materials experience a force is called a magnetic field.

- You can find the shape of a magnetic field using iron filings or plotting compasses.

- The Earth behaves as if it has a bar magnet inside it.

Electromagnets

Objectives

- Describe how to make an electromagnet
- Describe how to change the strength of an electromagnet

Switching magnets on and off

A bar magnet is a permanent magnet. You cannot switch it off without demagnetising it completely. You can make a magnet that you can switch on and off, using a coil of wire and a battery. It is called an **electromagnet**.

The field around a wire

Electromagnets work because when an electric current flows through a wire, a magnetic field is produced. You can use plotting compasses to draw the field around a single piece of wire.

The magnetic field lines around the wire are circles. The magnetic field gets weaker as you get further away from the wire.

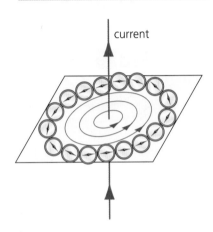

⬆ An electromagnet in the top of this toy keeps a metal Earth hovering below it.

Making an electromagnet

You can make a circular loop of wire and pass a current through it. The magnetic field lines at the centre are straight.

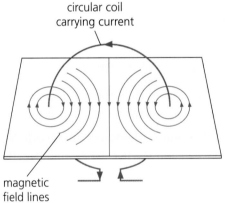

You can a coil a wire into lots of loops and make a coiled cylinder. When you pass an electric current through the coil, a magnetic field is produced. It has a similar shape to the field around a bar magnet. Coiling the wire concentrates the magnetic field inside the loops, so the field is much stronger than for a single loop on its own. A coil of wire is called a **solenoid**.

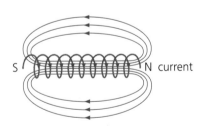

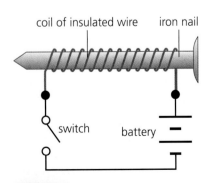

To make an electromagnet, you use a coil with some magnetic material called a **core** in the centre. The core becomes magnetised by the magnetic field of the coil. This makes the electromagnet much stronger. Most cores are made of iron.

The switch turns the electromagnet on and off because the magnetic field is only produced when a current is flowing in the wire. The **iron core** does not remain magnetised when you turn the electromagnet off.

The strength of an electromagnet

The strength of an electromagnet depends on:

- the number of turns, or loops, on the coil – more turns will make a stronger electromagnet
- the current flowing in the wire – more current will make a stronger electromagnet
- the type of core – using a magnetic material in the core will make a stronger electromagnet.

The current flowing in the wire depends on the power supply and the type of wire that is used in the electromagnet.

Permanent magnets and electromagnets

There are two main differences between a permanent magnet and an electromagnet.

- You can turn an electromagnet on and off.
- You can make an electromagnet that is much stronger than a permanent magnet.

This makes electromagnets very useful. They can be used to move large pieces of iron or steel in a factory, or to move cars in a scrapyard. When the current is turned off, the electromagnet drops the car.

Doctors can use a small electromagnet to remove iron or steel splinters from people's eyes.

1 Describe how you could use an electromagnet to sort a mixture of iron and copper pieces into two separate piles of iron and copper.

2 What do you think would happen to the magnetic field around the single piece of wire if the direction of the current was reversed?

3 Most electromagnets have a core.

  a Why is iron usually used for the core?

  b What would happen if you used steel instead?

4 A student says that an electromagnet is stronger when it has an iron core because the electric current flows through the iron as well as the wire. Is this correct? What would you say to the student?

- A current flowing through a wire produces a magnetic field.
- An electromagnet is a coil of wire usually wrapped around an iron core.
- You can switch an electromagnet on and off by turning the current on and off.
- The magnetic field around an electromagnet is similar to that around a bar magnet.

Identifying and controlling variables

Electromagnet investigation

Aanjay and Citra made an electromagnet by winding insulated wire around an iron nail. When they turn the current on, the wire produces a magnetic field that magnetises the nail.

They are experimenting with metal paper clips.

The number of paper clips tells us how strong the electromagnet is.

We could find out what affects how many paper clips the electromagnet can pick up.

Dependent and independent variables

The number of paper clips is the **dependent variable**. It is dependent because it will change when they make changes to the electromagnet.

In an investigation, the dependent variable is the thing that you measure after you make changes.

The thing that you make changes to is the **independent variable**. It is independent because you can choose how to vary it.

Aanjay and Citra think about the things that they could change. This is the list of variables that they think of first:

- the number of turns on the electromagnet
- the type of core
- the current in the wire.

They know that the current in the wire depends on the voltage of the power supply or battery, and the type of wire that they use to make the coil. This is their final list of variables:

- the number of turns on the electromagnet
- the type of core
- the type of wire
- the voltage of the power supply or battery.

We need to make sure that we only change one variable at a time.

Yes, we need to say how we have controlled all the other variables

Controlling variables

Aanjay and Citra decide that they will change the number of turns on the electromagnet. They write down how they will control the other variables.

Anjay's plan

We are going to change the number of turns on the coil and measure the number of paper clips. We will keep the type of core, the voltage of the battery, and the type of wire the same. Here is the table for my results.

Number of turns	Type of core	Voltage of battery	Type of wire	Number of paper clips
5	iron	6 V	copper	
10	iron	6 V	copper	
15	iron	6 V	copper	
20	iron	6 V	copper	
25	iron	6 V	copper	
30	iron	6 V	copper	

Citra's plan

We are going to change the number of turns on the coil and measure the number of paper clips. We will keep everything else the same. Here are the values for the variables that we are keeping the same:

- type of core = iron
- voltage of the battery = 6 V
- type of wire = copper

Here is the table for my results.

Number of coils	Number of paper clips
5	
10	
15	
20	
25	
30	

In both plans it is clear which variable is the independent variable, which variable is the dependent variable, and how the other variables will be controlled during the investigation.

There are lots of other things that could change while the students are doing their investigation. They only need to control them if they will affect the strength of the electromagnet.

Q

1 What is the difference between a dependent variable and an independent variable?

2 Eshe is investigating the link between the extension of an elastic band and the force that she applies. She changes the force and measures the extension.

   a Which is the independent variable?

   b Which is the dependent variable?

3 A group of students plan to investigate how high a ball bounces on different surfaces. This is part of their plan.

   a Which is the independent variable?

   b Which is the dependent variable?

   c Describe one way in which their plan could be improved.

   d Draw the table of results that you would need for this investigation.

We are going to change the surface and measure the height of the bounce.

We are going to keep everything else the same.

We will use a tennis ball each time and drop it from the same height.

- Independent variables are things that you choose to change.
- Dependent variables are things that change as a result.
- In an investigation you only change one variable at a time.
- You must clearly state how you have controlled the other variables.

Using electromagnets

Objectives

- Describe some uses of electromagnets
- Explain why electromagnets are used instead of permanent magnets

Switching off

An X-ray machine can be very dangerous. The X-rays can damage your cells if you are exposed to them for too long. The radiographer, who operates the machine, should not be near the machine every time it is switched on and off.

The machine uses a high-voltage circuit. The radiographer turns it on and off using a **relay** instead of a normal on/off switch.

The relay

A relay uses a small current in one circuit to operate a switch in a separate high-voltage or high-current circuit.

You can make a simple relay with a **reed switch**. This is made of two pieces of magnetic material (called reeds) with a gap between them. If you wind a coil of wire around the outside and pass a current through it, the two reeds are magnetised. They attract each other and touch. If the reeds are part of another circuit, then current will flow in the second circuit when they touch.

A relay is often used to turn on the starter motor in a car. The starter motor works from the car battery, which produces a dangerously high current. The relay circuit that the driver switches on has only a small current, but it turns on the high-current starter motor circuit.

reed switch
circuit 1   circuit 2
magnetic reeds

⬆ If you close the switch in circuit 1 the lamp in circuit 2 will come on.

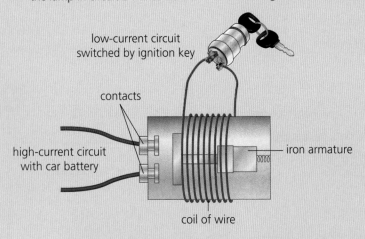

low-current circuit switched by ignition key

contacts

high-current circuit with car battery

iron armature

coil of wire

When the driver turns the key, a current flows in the coil of wire which produces a magnetic field round it. This magnetises the iron **armature**. The armature is attracted towards the contacts. It connects the contacts and completes the starter motor circuit so that the battery powers the starter motor. It only takes a small current to operate the electromagnet. The current in the starter motor circuit is much bigger and could be dangerous if the driver had to switch it on directly.

Fire doors

In big buildings, fire doors are important to stop a fire spreading. But going through lots of fire doors can be a nuisance when there is no fire.

A fire door can be held open by an electromagnetic switch on the wall. While there is no fire, a current flows in the coil of the electromagnet. A piece of iron on the door is attracted to the electromagnet on the wall, holding the door open. When the fire alarm is triggered, the circuit is broken so there is no current flowing in the coil of the electromagnet. The door closes.

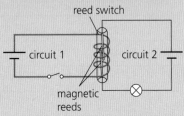

The electric bell

Some doorbells contain an electromagnet. In some ways they operate like a relay, but with an important difference.

When a visitor presses the bell-switch, the circuit is complete.

- A current passes through the coil and the iron core is magnetised.
- The iron armature is attracted to the core and the hammer hits the gong, making a sound.
- This breaks the circuit at X so the iron core is no longer magnetised.
- The armature springs back and completes the circuit.
- The sequence starts again so the bell keeps ringing for as long as the visitor presses the bell.

Unlike in the relay, the armature moves backwards and forwards while the switch is pressed.

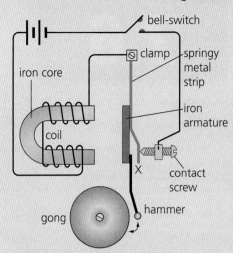

Using magnets in medicine

Magnets can be used to produce an image of the inside of your body. The device is called a **magnetic resonance imaging (MRI) scanner**. An MRI scanner contains very strong magnets. These produce a very intense magnetic field that interacts with the hydrogen in the water molecules in your body. The MRI scanner enables doctors to take pictures of people's organs.

MRI scans are safe, painless, and produce images that are much more detailed than X-rays. Doctors can use them to see structures inside the body that X-rays don't show, without having to cut the patient open. MRI scanners allow doctors to see how an organ works in normal conditions. Unfortunately MRI scanners are very expensive, and the patient has to lie very still in a small space for the scan to be taken.

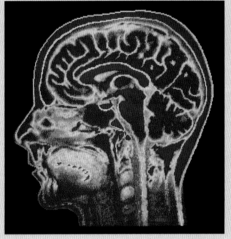

1. In what situations might you need to use a relay rather than a normal switch?

2. Look at the diagrams of the relay and the doorbell.

   a. Explain why the doorbell circuit is called a 'make-and-break' circuit.

   b. How is the doorbell circuit different from the relay circuit?

   c. How is it the same?

3. How could you use an electromagnet to make a burglar alarm? (*Hint:* think about how the fire doors work. You may want to draw a picture to help to explain how it works.)

4. Make a list of the pros and cons of MRI scans.

- A relay is a type of switch that can be used in a low-current circuit to turn on a high-voltage or high-current circuit.

- Fire doors use electromagnets to close automatically.

- An electric bell works by continuously making and breaking a circuit.

- Scanners use magnets to take pictures of organs in the body.

1 For each of these statements about magnetic materials, write 'true' or 'false'.

   a Magnetic materials will attract other magnetic materials.

   b Magnetic materials will be attracted to magnets.

   c Magnetic materials will be repelled by magnets.

   d Magnetic materials will repel other magnetic materials. [4]

2 You have a small piece of each of the materials listed below. Which of them would be picked up by a magnet?

   A nickel

   B steel

   C iron

   D copper

   E zinc

   F cobalt

   G magnesium [1]

3 A student is playing with two magnets on the desk. She puts a blue magnet on the desk and brings a green magnet near to it. When she does this the blue magnet moves.

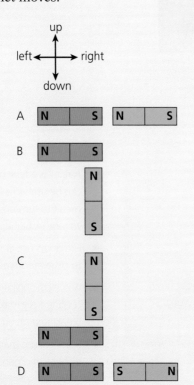

a For each diagram, write in which direction the *blue* magnet will move, using the words up, down, left, or right. [4]

b There is one direction that you have not used to answer part **a**. Draw a diagram of two magnets to show how you could make the blue magnet move in that direction. [1]

4 Copy and complete these sentences. You may need to use the same word or phrase more than once.

   a We can draw _____ _____ lines to show the pattern of the _____ _____ around a bar magnet. [1]

   b If there are lots of lines close together, that means that the _____ _____ is very _____. This happens near the _____ of the magnet. [1]

   c All magnets have a north and a south _____. [1]

5 Icha wants to investigate the magnetic field around two magnets.

   a Name two methods that she could use to find out about the shapes of the magnetic fields. [1]

She puts two magnets on the table so that they are repelling.

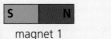 X

   magnet 1   magnet 2

   b X is the neutral point. What is mean by 'neutral point'? [1]

   c Which magnet is stronger? Explain your answer. [1]

   d Apart from the two methods you have described in part **a**, describe a method that you can use to find the position of the neutral point. [1]

6 Here is a diagram of a plotting compass when there is no magnet near it and when it is near the end of a bar magnet.

A compass with no   B compass near the end
  magnet near it   of a bar magnet

Explain what is happening to the needle in each picture. [2]

7 Explain what is meant by:

  a a permanent magnet [1]

  b a magnetic material [1]

  c a magnetic field [1]

  d magnetic field lines [1]

  e an electromagnet. [1]

8 The Earth has a magnetic field shaped as if it had a bar magnet through its centre.

  a Copy the diagram of the Earth with a bar magnet at the centre and label the poles of the imaginary magnet. [1]

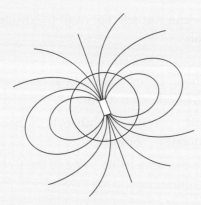

  b Explain your choice of labels. [1]

  c Add arrows to the magnetic field lines. [1]

  d A compass needle lines up with the Earth's magnetic field. What would happen if you were trying to use a compass near a large piece of lodestone? [1]

9 Abasi wants to make an electromagnet.

  a Describe how he could make a simple electromagnet using a piece of wire, a nail, a battery, and a switch. [2]

  b He wants the electromagnet to be strong. What could he do to increase the strength of the electromagnet? [2]

  c He and his friends make four different electromagnets.

    Put the electromagnets below in order of strength, starting with the weakest. Write the letters in the correct order.

    A 50 turns, low current, copper core

    B 100 turns, high current, iron core

    C 50 turns, low current, iron core

    D 100 turns, low current, iron core

10 Which of these statements are true? There may be more than one. [1]

  a You can make a magnetic material into a magnet by stroking it with another magnet.

  b You cannot change the strength of an electromagnet.

  c A compass points to the North Pole, at the top of the axis on which the Earth spins.

  d You can demagnetise a magnet by hitting it with a hammer.

  e Iron is hard to magnetise.

11 There are lots of different metals that can be used to make the core of an electromagnet. A student wants to use a metal that will be magnetised while a current flows in the wire of the electromagnet, but will not stay magnetised when the current is switched off.

  a Choose the correct metal or metals that could be used from the list below. [1]

    iron or steel   iron   copper   copper or iron   steel

  b Explain why the other metal or metals would not be suitable. [1]

12 Ikenna is going to investigate how the type of core affects the strength of an electromagnet.

  a Which is the independent variable? [1]

  b Which is the dependent variable? [1]

  c Which variables will Ikenna need to control? [1]

  d He has a choice of small paper clips, large paper clips, or iron filings to measure the strength of the electromagnet. Which should he choose? Explain your answer. [1]

  e Draw a table that he could use for his results. [1]

1 Look at this distance–time graph for a cyclist.

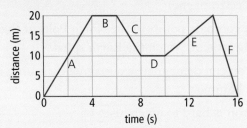

   a Calculate the speed during each of the sections of the graph A–F. [2]

   b Explain why the graph is not very realistic. [1]

2 Copy and complete this table to show the similarities and differences between light and sound. [8]

Statement	Applies to light only	Applies to sound only	Applies to both light and sound
only travels in a straight line			
can travel through a vacuum			
travels faster through liquids than gases			
travels as waves			
travels faster through gases than solids or liquids			
transfers energy			
is detected by our ears			
is detected by our eyes			

3 The first person to stand on the Moon took 3 days to get to the Moon from the Earth. The distance to the Moon is 384 400 km.

   a Calculate the average speed of the spacecraft. [1]

   b Explain why the speed that you have calculated is the *average* time. [1]

   Here is distance–time graph for the first 10 seconds of the launch.

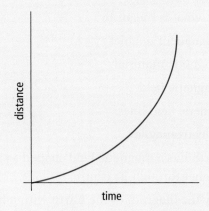

   c What is happening to the rocket during this time? [1]

   d How can you tell that from the distance-time graph? [1]

4 A student writes a story about a child's first day at a new school. Unfortunately he is confused about the use of the following words:

   **transmit   reflect   absorb   transparent
   translucent   opaque**

   Rewrite the story using some of the words in the list above. [6]

   One day Rajiv was walking to school. He could see the people in their cars through the windows because they were opaque. He could not see the lorry around the corner through the transparent wall. He wondered if his friend was in the shop. The light came through the opaque window but he could not see in.

5 Here is a diagram showing a ray of light hitting a glass pane in a window.

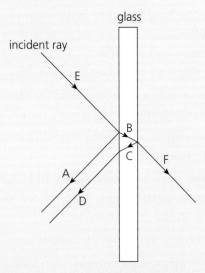

   a Give the letters of two reflected rays. [1]

   b Give the letters of two refracted rays. [1]

   c The incident ray is brighter than ray A. Why? [1]

If you look out of a window in the daytime you see straight through it. If you look through it at night you see your reflection in the window.

d Explain why you don't see a reflection during the day. [1]

e Explain why you can see your reflection at night. [1]

f Explain why you can see the colour of the clothes that you are wearing when you look in the mirror. [1]

g Why is it harder to see colour of your clothes in the reflection of a window at night? [1]

6 The speed of light is 300 000 km/s.

a How far does light travel in 10 seconds? [1]

b How far is 10 light seconds? [1]

c Explain why you do not need to do a calculation to answer part **b.** [1]

d List these speeds in order from fastest to slowest: [1]

A the speed of light in glass

B the speed of light in air

C the speed of sound.

7 A student is planning an investigation into the effect of changing the tension in a string on the pitch of the sound produced. Here is a diagram of the experiment that he is planning to do.

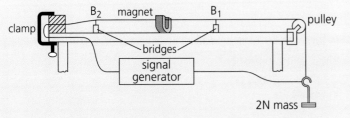

He plans to change the masses hanging on the wire and look at the wave on the screen of the oscilloscope. He can use the waves on the screen to calculate the frequency.

a What is the independent variable? [1]

b What is the dependent variable? [1]

c What variable or variables will he need to change? [1]

d Draw a table he could use to record his results. [1]

e What sort of graph could he plot of his results? Explain your answer. [1]

f He does the experiment and finds that as he adds more masses the frequency gets bigger. Name a musical instrument in which this would be important. [1]

8 A referee needs to blow a whistle in a football game to stop play. This is a picture of the wave on the screen of the oscilloscope:

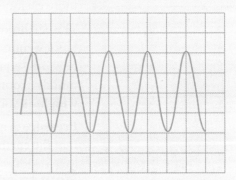

He blows his whistle again, but this time with a lower pitch and softer.

a Sketch a wave to show what this sound would look like. [1]

b Explain why you have drawn the wave in the way that you have. [1]

9 Rani has 3 rods that look the same. One is made of iron, one is made of copper and one is a magnet.

a Describe how she could use a magnet to work out which is which. [1]

b Her brother wants to make an electromagnet with some wire and a battery. Explain how to make an electromagnet. [1]

c Which rod, if any, should he use for the core? Explain your answer. [1]

d What will Rani notice if she moves a compass around the magnetic rod and around her brother's electromagnet? Why? [1]

Rani and her brother are watching a television programme about cars. When cars are scrapped they are picked up by an electromagnet.

e Describe three differences between the electromagnet used in a scrapyard and an electromagnet you use in class. [1]

8.1 Pressure

Objectives

- Explain the difference between weight and pressure
- Be able to calculate pressure

What is pressure?

It can be very difficult to walk across mud or sand without getting stuck. Earthmovers use wide tracks instead of ordinary tyres so that they can move across the mud more easily.

The earthmover is very heavy. It has a weight of about a million newtons, equal to about 15 000 people! A single person standing on the same muddy ground might sink. But the earthmover does not sink because its **weight** is spread out over a bigger area.

Pressure is a measure of how much force is applied over a certain area. We say that the earthmover produces a smaller pressure on the mud than the person.

How to calculate pressure

You can calculate pressure using this equation:

$$\text{Pressure} = \frac{\text{force}}{\text{area}}$$

Force is measured in newtons (N) and area is measured in metres squared (m²), so pressure is measured in newtons per metre squared (N/m²). 1 N/m² is also called 1 **pascal** (Pa).

Sometimes it is easier to measure smaller areas in centimetres squared, cm². If you measure the area in cm² and the force in N, then the pressure is measured in N/cm².

A force of 100 N is spread over an area of 2 m². What is the pressure?

$$\text{Pressure} = \frac{\text{force}}{\text{area}}$$
$$= \frac{100 \text{ N}}{2 \text{ m}^2}$$
$$= 50 \text{ N/m}^2$$

A force of 20 N is spread over an area of 4 cm². What is the pressure?

$$\text{Pressure} = \frac{\text{force}}{\text{area}}$$
$$= \frac{20 \text{ N}}{4 \text{ cm}^2}$$
$$= 5 \text{ N/cm}^2$$

When you do calculations it is very important to look at the units of area. If you write them next to the number in your equation, then you will see which unit of pressure you need to use.

Large and small pressure

Sometimes it is useful to make the pressure *bigger*.

The studs on a football boot have a small area compared with the area of the foot. This produces a bigger pressure, so that the studs sink into the ground and help the footballer to grip the ground and move more easily.

The weight of a footballer is 800 N.

The area of his two feet is 200 cm².

$$\text{Pressure} = \frac{\text{force}}{\text{area}}$$

$$= \frac{800 \text{ N}}{200 \text{ cm}^2}$$

$$= 4 \text{ N/cm}^2$$

The total area of the studs is 20 cm².

$$\text{Pressure} = \frac{\text{force}}{\text{area}}$$

$$= \frac{800 \text{ N}}{20 \text{ cm}^2}$$

$$= 40 \text{ N/cm}^2$$

The same force over a smaller area produces a much bigger pressure.

At other times it is useful to make the pressure smaller.

The weight of the earthmover is 1 000 000 N.

The total area of four normal tyres is 2 m².

$$\text{Pressure} = \frac{\text{force}}{\text{area}}$$

$$= \frac{1\,000\,000 \text{ N}}{2 \text{ m}^2}$$

$$= 500\,000 \text{ N/m}^2$$

The total area of the earthmover tracks is 25 m².

$$\text{Pressure} = \frac{\text{force}}{\text{area}}$$

$$= \frac{1\,000\,000 \text{ N}}{25 \text{ m}^2}$$

$$= 40\,000 \text{ N/m}^2$$

The same force over a bigger area produces a much smaller pressure. A pressure of 40 000 N/m² is the same as a pressure of 4 N/cm², much smaller than the studs.

You can use the equation for pressure to calculate the force or the area.

$$\text{Pressure} = \frac{\text{force}}{\text{area}} \qquad \text{Force} = \text{pressure} \times \text{area} \qquad \text{Area} = \frac{\text{force}}{\text{pressure}}$$

1 Copy and complete this table.
Don't forget the units!

Force	Area	Pressure
150 N	25 cm²	
60 N	15 m²	
5 N	0.1 cm²	

2 A pressure of 0.01 N/cm² is produced when a book with an area of 300 cm² is placed on a table. What is the weight of the book?

3 A brick has a weight of 15 N. If the pressure it produces is 0.5 N/cm² what is the area in contact with the ground?

4 A hippopotamus produces a pressure of 250 000 Pa when it is standing on all four feet. If the weight of the hippopotamus is 40 000 N, what is the area of each foot?

- Pressure = $\dfrac{\text{force}}{\text{area}}$
- Pressure is measured in N/m² or N/cm².
- 1 N/m² is the same as 1 Pa (pascal)
- A small force over a big area produces a small pressure.
- A big force over a small area produces a big pressure.

The effects of pressure

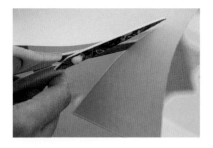

Scissors, knives, and tools

The blades of knives and scissors are sharp. They have a very small area so the force that you apply produces a big pressure.

If the blade becomes blunt then the area gets bigger. Now the pressure might not be big enough to cut things. It is important to keep knives and scissors sharp.

You can use a hammer to hammer a nail into a piece of wood. The end of the nail is pointed, so it has a very small area. The pressure produced on the wood when you hit the nail with the hammer will be very large.

Lots of tools have sharp edges so that they can cut or make holes in things easily.

Walking on mud

Some animals need to travel across mud, sand, or other soft surfaces.

Wading birds have wide flat feet to reduce the pressure on the ground. This means that they do not sink so much when they have to walk over a soft surface.

Sandy feet

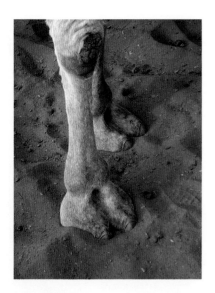

It is much easier for camels to walk across sand than it is for horses to walk across sand. The area of a camel's foot is much larger than the area of a horse's hoof.

The weight of a camel is 5000 N.

The area of its four hooves is 2000 cm².

$$\text{Pressure} = \frac{\text{force}}{\text{area}}$$
$$= \frac{5000 \text{ N}}{2000 \text{ cm}^2}$$
$$= 2.5 \text{ N/cm}^2$$

The weight of a horse is 4000 N.

The area of its four hooves is 400 cm².

$$\text{Pressure} = \frac{\text{force}}{\text{area}}$$
$$= \frac{4000 \text{ N}}{400 \text{ cm}^2}$$
$$= 10 \text{ N/cm}^2$$

Even though a horse has a smaller weight than a camel, the pressure that it produces on the ground is four times bigger because the area of its hooves is smaller.

Quicksand

Some people are worried that they could be caught in quicksand. Quicksand does not pull you in, but it is difficult to walk across it. Why can you walk across sand but not across quicksand?

Quicksand is a mixture of sand and water found close to beaches, swamps, and marshes, but not usually found in the desert. It appears to be solid, but if you stand on it the pressure that you produce on it means that you will sink into it. This is because the water reduces the friction between the grains of sand so your weight can push the sand grains out of the way, much more easily than with dry sand. For someone who strays into quicksand, if they lie down and do not struggle they should be able to float!

Wet sand near the edge of the sea contains less water than quicksand. It is more like a solid. If you walk on wet sand the pressure that you produce forces the water out of the gaps between the grains of sand. There is more friction between the grains so it feels like you are walking on solid ground.

Pressure and bikes

Some bikes are designed to travel across mud and others are designed to be used on the road. The tyres of road bikes are much narrower than those of an off-road bike.

Q

1 Explain why the feet of camels and wading birds have the shapes that they have.

2 Two birds have the same mass but one has much bigger feet. How would the pressure that they produce on the ground be different?

3 What would happen if you tried using a road bike to ride across a soft surface? Explain your answer.

4 Two horses have hooves of about the same area. Horse A has a much bigger mass than horse B. How would the pressure that they produce on the ground be different?

- Increasing the area reduces the pressure, which can be useful in some situations.
- Reducing the area can be useful if you need a large pressure.

Pressure in liquids

Liquid pressure

If you go swimming you can feel the pressure of the water. Your ears detect **liquid pressure**. It doesn't matter how you move your head underwater you still detect the same pressure.

Pressure in liquids and gases is different from the pressure produced when one solid is on top of another. Liquid pressure is due to the forces between the liquid particles and the surfaces of a container, like the walls of a swimming pool.

The particles in a liquid are very close together so liquids cannot be **compressed**. We say that they are **incompressible**.

Isha fills a syringe full of water. He holds one finger over the hole so the water cannot get out. Then he presses down on the plunger of the syringe. Nothing happens! It is not possible to force the liquid into a smaller space.

If you apply a force to a liquid the forces between the particles increases, and the forces act in all directions. This means that liquid pressure acts in all directions.

You can demonstrate that the pressure in a liquid acts in all directions. In the bag, the water comes straight out of each hole and then falls because of gravity. The liquid pressure pushes the water out of the holes – it doesn't act just downwards like gravity does.

It is liquid pressure that produces **upthrust**, the force that keeps things afloat.

Liquid pressure and depth

The pressure on a fish at the bottom of a lake is bigger than the pressure on a fish half-way down. At the bottom of a lake there is twice the weight of water pushing down compared with half-way down. The pressure gets bigger as you go down.

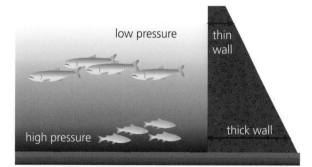

low pressure

thin wall

high pressure

thick wall

This is why the concrete at the bottom of a dam wall has to be many times thicker than the concrete at the top – because the pressure at the bottom is much, much bigger.

Pressure and diving

Fish that live very deep in the ocean can withstand very big pressures. When people go deep underwater they use diving vessels or submarines with very strong walls to withstand the pressure.

Divers need to understand about water pressure. They have to be careful how long they spend in very deep water because the nitrogen from the air that they are breathing dissolves in their blood. If they return to the surface quickly, this nitrogen forms bubbles as the pressure is reduced. The same thing happens when you open a can of carbonated drink. Lots of carbon dioxide bubbles form as the pressure is reduced.

Bubbles of nitrogen in a diver's blood are very dangerous and can cause great pain. The condition is called '**the bends**'. Deep-sea divers must return to the surface very slowly to stop the bubbles forming.

Measuring pressure

The pressure in a liquid or gas is measured with a **pressure gauge**. This contains a tube that is curled up. The tube straightens out as the pressure inside the tube increases. This moves a needle to show a reading on a scale.

needle   coiled metal tube

pressure being measured

Q

1 What produces pressure in a liquid?

2 Why is the pressure at the bottom of the Pacific Ocean much bigger than near the surface?

3 A student fills a plastic bottle full of water. He make a hole in the bottle half-way up, and another hole at the bottom.

 a Does the water come faster out of the hole in the middle or the hole at the bottom? Explain your answer

 b The water comes out at right angles to the surface of the bottle. Why?

 c As the water drains out what happens to the speed of the water leaving both holes? Why?

!

- There is pressure in a liquid because of the forces between particles in the liquid acting over an area.

- As you go deeper the pressure gets bigger because there is more weight of water above you.

- You can measure liquid pressure with a pressure gauge.

Using pressure in liquids

How do hydraulic machines work?

Hydraulic machines use liquids to produce very large forces. A hydraulic jack is used to lift up a car so the mechanic can repair it.

In a gas there is lots of space between the particles, so you can compress a gas. Hydraulic machines work because unlike a gas, you cannot squash a liquid – liquids are incompressible. The particles in a liquid are effectively touching each other, so liquids pass on any pressure applied to them.

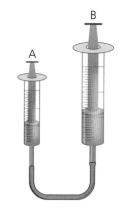

The two syringes connected together show how a hydraulic jack works. What will happen if you push down on plunger A? Plunger B will move up.

The force that you apply when you push down on A produces pressure in the liquid. This pressure is passed through the liquid to B. It produces a force which pushes B up.

Increasing the force

A hydraulic machine can be used to increase the size of the force.

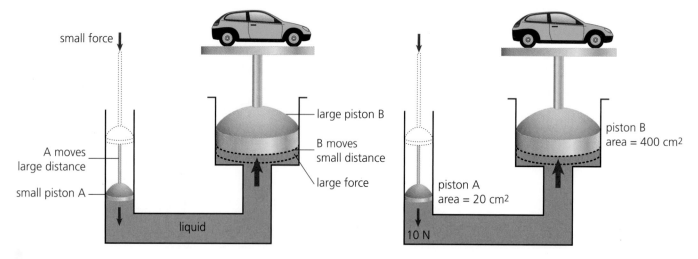

The left-hand diagram shows how a hydraulic jack works. Two metal pistons are connected by a tube of liquid. The pistons can move in and out of the cylinders like the plunger in a syringe. When the mechanic applies a force to piston A, piston B will go up. The jack allows the mechanic to lift the car.

If you push down with a force of 10 N on piston A, as shown in the right-hand diagram, then a pressure is produced in the liquid that is passed through the liquid to piston B.

$$\text{Pressure} = \frac{\text{force}}{\text{area}}$$

$$= \frac{10 \text{ N}}{20 \text{ cm}^2}$$

$$= 0.5 \text{ N/cm}^2$$

The liquid transmits this same pressure to piston B. What is the force at B?

Force = pressure × area

    = 0.5 N/cm² × 400 cm²

    = 200 N

The force has increased from 10 N at A to 200 N at B. Because B has a bigger area, the force that is produced at B is bigger. Piston B is 20 times bigger so the force is 20 times bigger.

A hydraulic jack is called a **force multiplier** – the force at A has been multiplied. You might think that you are getting a bigger force for nothing, but this is not the case. A hydraulic jack will produce a bigger force, but the larger piston will move a much smaller distance than the smaller piston. You can have a bigger force but it will not move the piston as far. In a real pump, the mechanic has to pump more oil in to lift the car.

Hydraulic brakes

Most cars use **hydraulic brakes**. It would be impossible for the driver to produce a force big enough to stop the car. So the hydraulic braking system makes the driver's force bigger.

It is very important for the braking system to work properly. If air gets into the brake fluid, the brakes will not work properly.

Hydraulic press

A machine called a **hydraulic press** can produce a force big enough to shape a sheet of metal to make a car door. Hydraulic presses are also used to crush cars.

⬆ Hydraulic machines multiply our force to do jobs that we could not do on our own.

Q

1 Explain how a hydraulic jack can be used to lift a car.

2 Why are hydraulic brakes less effective if there is air in the liquid?

3 Copy and complete the table below for a simple hydraulic machine.

Area of piston A	Force applied to piston A	Pressure in the liquid	Area of piston B	Force produced by piston B
2 cm²	8 N		25 cm²	
5 cm²	20 N		20 cm²	
0.01 m²	6 N		0.1 cm²	

4 A student pushes down with a force of 2 N on the plunger of syringe A. Syringe B has a plunger with an area that is 3 times the area of plunger A. What is the force produced by plunger B?

!

- Hydraulic machines use liquids, which are incompressible.
- Hydraulic machines make forces bigger.
- Hydraulics are used in brakes, for crushing cars, and for making cars.

Pressure in gases

Gas pressure

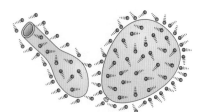

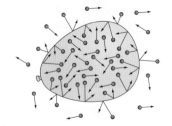

Raj blows up a balloon. The balloon gets bigger and bigger. Suddenly it bursts! Why does this happen?

Raj pumps more air particles into the balloon. They bump into each other and into the rubber of the balloon and push the balloon outwards. The collisions produce **gas pressure**. By pumping more air into the balloon, Raj increases the pressure inside it.

Air particles hit the outside of the balloon too, but as the pressure inside increases there are more collisions on the inside than on the outside. When the gas pressure inside is really big, the balloon bursts.

A bicycle tyre does not keep on expanding like a balloon does. Pumping up a bicycle tyre increases the gas pressure inside the tyre.

In a gas there are big gaps between the particles. As you pump more and more gas into a closed space like a tyre, the particles are pushed closer together – the gas becomes *compressed*. There are more particles so there are more collisions, producing a bigger pressure.

What is the link between pressure and volume?

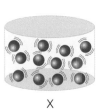

X

Y

Both these boxes contain the same number of gas particles. If box X has half the volume of box Y, there will be twice as many collisions between the gas particles and the walls of box X. This means that the pressure in X will be double the pressure in Y.

As the volume goes down, the pressure goes up. The opposite is also true. If the volume increases, then the pressure will decrease.

Relationships

We use the words '**directly proportional**' to describe the relationship between two things that increase together or decrease together in a regular way. For example, when you put weights on a spring, if you double the force then the extension will double. The extension is directly proportional to the force.

With gases, the pressure and volume are **inversely proportional**. As one quantity goes up, the other goes down. If you double the volume of a gas, the pressure will *halve*. If you halve the volume of a gas, the pressure will double.

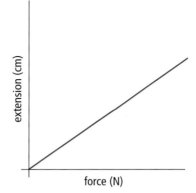

⬆ If we plot a graph of two quantities that are directly proportional, the graph is a straight line through (0,0).

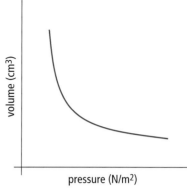

⬆ If we plot a graph of two quantities that are inversely proportional, the graph is a curved line.

Atmospheric pressure

The air around us exerts gas pressure on our bodies all the time. This is called **atmospheric pressure**. We do not feel the pressure because it is balanced by the pressure of the gases and liquids in our bodies pushing outwards.

The atmospheric pressure at sea level is bigger than the atmospheric pressure high up a mountain. Gravity pulls the air particles towards the Earth. The air near the Earth's surface is compressed, so the air near the surface of the Earth. If the particles are closer together, then the pressure is bigger. This means that there will be more collisions producing a bigger force and a bigger pressure.

Some mountain climbers carry oxygen tanks when they climb high mountains such as Mount Everest. There is less oxygen high up. The tanks contain oxygen gas that has been compressed into a small volume. The gas has to be compressed because otherwise it would not be possible for a climber to carry the number of containers of oxygen that he would need. The pressure inside the oxygen tank is very high. The tanks are made of metal which can withstand the high pressures. People who do not take oxygen can experience altitude sickness and nosebleeds.

Q

1 What causes gas pressure?

2 Why can you compress a gas but you cannot compress a liquid?

3 Why does the pressure inside a bicycle tyre increase as you pump it up?

4 Aircraft cabins are 'pressurised'. Air is gradually pumped into the aircraft as it climbs, and then let out again when it lands. Why?

5 **Extension:** what would happen to the pressure in a gas if you compressed it into a volume that was one-third the size? Why?

- Gas pressure is caused by particles colliding with the walls of the container.
- If you compress a gas into a smaller volume, the pressure will be bigger.

Pressure, volume, and temperature in gases

Which tyres?

In Formula 1 racing, the choice of tyres is extremely important. The teams have to work out which type of tyre to use for different weather conditions.

It is also important to keep the tyres at the right temperature. Sometimes the mechanics use tyre warmers to keep them warm ready for a race.

Pressure in tyres

The pressure inside a tyre changes with the **temperature**. If you are cycling on a hot day, the pressure inside the tyres will be bigger than on a cold day.

To understand why the pressure is bigger, think about the air particles. In a gas the particles move around very fast. When they collide with the walls of a container, like the inside of a tyre, they produce a pressure on the walls that keeps the tyre inflated.

If you heat a gas, its particles move faster. If the gas is trapped in a container that cannot get any bigger, like a tyre, the gas pressure gets bigger. This is because there are more collisions between the gas particles and the container walls.

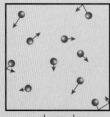

cool gas, less
energetic, fewer
collisions

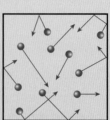

hot gas, more
energetic, more
collisions

What is the link between pressure and temperature?

Here are some data from a company that makes tyres. The graph shows how the tyre pressure changes with temperature.

The graph shows that as the temperature rises, the pressure increases. For each increase of 20 °C, the tyre pressure increases by the same amount.

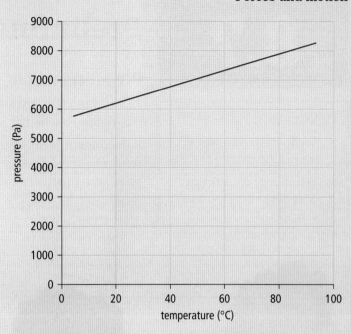

What is the link between volume and temperature?

Isaac knew that when a gas is in a container that cannot expand, the pressure is inversely proportional to its volume. If you double the volume the pressure will halve. He wanted to find out what happens to the volume when you heat a gas that *can* expand.

He took a balloon and blew it up to less than half full. He measured the circumference of the balloon by putting a piece of string around the widest part of the balloon.

Next Isaac held the balloon in hot water, being careful not to hurt his hands. He took the balloon out of the water and measured the circumference. Finally he held the balloon in very cold water, and measured the circumference again.

The table shows Isaac's results.

Temperature of water	Circumference (cm)
hot	18
warm	17
cold	16

Isaac found out that as the temperature rises, the volume of the balloon increases. The volume of the balloon expands until the pressure inside and outside the balloon is the same.

1 Copy these sentences and fill in the gaps.

If you increase the temperature of gas in a container that cannot expand, the pressure will _____. This is because the particles in the gas are moving _____ and colliding _____ often·

2 Explain the results of Isaac's experiment using the idea of air particles.

3 a How could Isaac change his experiment so that he could plot a line graph?

b What would you expect his line graph to look like?

4 If a bicycle tyre gets worn thin, you might get a puncture. Is this more likely to happen on a hot or a cold day? Explain your answer.

- If the temperature of a gas increases, its pressure increases.
- If the temperature of a gas increases, its volume increases
- You can explain observations of pressure, temperature, and volume using a particle model.

Preliminary work

Preparing a plan

Jabari wanted to investigate what happens when you apply pressure to ice. He wondered whether ice gets harder, or softer, or even melts.

Sometimes it is difficult to know how to start an investigation. You might think of a way of doing the investigation, but not be confident that it will work. You might not be sure what equipment would be best to use, or what results to take.

Objectives

- Know what is meant by preliminary work
- Understand how preliminary work can be used to plan an investigation

No, the preliminary work helps me to work out how to do it.

But aren't you going do the investigation twice?

Jabari wasn't sure how to go about his investigation. He decided to do some **preliminary work** to find out how to do it. His friend Zane was confused about what preliminary work is for.

Jabari explained to Zane that there are lots of reasons to do preliminary work. Preliminary work can help you to decide:

- what method to use
- what equipment to use
- how many results to take
- the range of results to take
- how to get accurate and reliable results
- how to work safely.

Trying things out

Jabari took some ice cubes out of the freezer and put them on a plate. He tried pressing down on the ice with a metal coin but the ice cube kept slipping from underneath the coin.

I need bigger blocks of ice to do my investigation.

Jabari made some bigger blocks of ice by putting trays of water in the freezer. He took out a block of ice and put it on a big tray. He tried pushing the coin down again.

How am I going to measure the force? I need to hang weights on it so that I know the size of the force.

Jabari took the block of ice and supported it on two pieces of wood.

What can I use to hang my weights on?

He collected some string and some weights and hung them over the ice. He made sure that he wore safety goggles.

Jabari waited but nothing happened. He wondered if the experiment would work with wire. He found that the wire started to cut through the ice, but it took much longer than he expected.

Maybe you need thinner wire?

Maybe I need more weight? Or perhaps both?

Doing the experiment

Jabari took a thinner wire and hung lots of weights on it. He found that now the wire cut all the way through the block in a reasonable time.

He decided to investigate the link between the number of weights and the time that it took the wire to cut through the ice. His preliminary work had shown him how to design the investigation to get some results. He knew that he would need to start with a lot of weight to begin with, otherwise the experiment would take too long. He designed this table for his results:

Weight hanging from wire (N)	Time taken for wire to cut through the block (minutes)
20	
25	
30	
35	
40	
45	
50	

1 Make a list of the changes that Jabari made as a result of his preliminary work.

2 Why did Jabari have to wear safety goggles?

3 Why might it be difficult for Jabari to get precise results?

4 A student thinks that you should plan your investigation and not waste time doing preliminary work. What would you say to her?

Preliminary work can show you:

- how to get results in an investigation
- how many results to take and over what range
- how to work safely.

Density

What is density?

Density tells you how much **mass** there is in a certain **volume**.

These blocks all have the same volume, but they have different masses. This is because they have different densities.

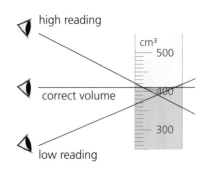
metal   wood   plastic

To work out the density of any material, whether solid, liquid, or gas, you need to measure two things: the volume and the mass.

Measuring the volume and mass of a liquid

You can find the volume of a liquid such as water with a **measuring cylinder.** Measuring cylinders measure volume in cm³ or millilitres (ml). 1 ml is equal to 1 cm³.

⬆ The volume of water is 400 cm³.

Water clings to the sides of its container, and curves upwards slightly at the edges. The curved shape is called a **meniscus**. You measure the reading at the bottom of the meniscus.

When you read the volume it is important to look straight ahead at the scale to get an accurate reading, rather than looking up or down at it.

high reading

correct volume

low reading

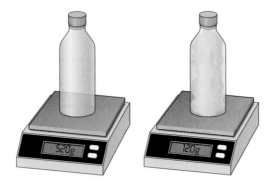

Once you know the volume, you need to measure the mass of the water to find its density. To measure the mass of a liquid you find the mass of the container first. You then pour in the water and measure the mass again. You can work out the mass of the water alone by subtracting the mass of the empty cylinder.

Measuring the volume and mass of a solid

Finding the volume of a regularly-shaped solid object is easy.

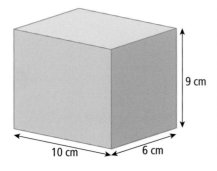

9 cm

10 cm   6 cm

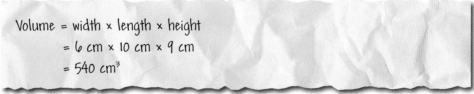

Volume = width × length × height
       = 6 cm × 10 cm × 9 cm
       = 540 cm³

If the solid does not have a regular shape, then you can find its volume by putting it in water, as shown on the right.

You measure the mass of a solid using a balance.

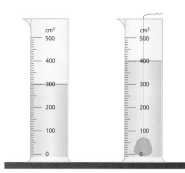

⬆ The volume of the solid is 100 cm³.

Calculating density

To calculate the density you use this equation:

$$\text{Density} = \frac{\text{mass}}{\text{volume}}$$

Mass is usually measured in kilograms (kg), but smaller masses are measured in grams (g). Volume is usually measured in metres cubed (m³), but smaller volumes are measured in centimetres cubed (cm³). The units of density will depend on the units of mass and volume that you use. Here are some examples.

The block of wood has a mass of 270 g and a volume of 540 cm³. $$\text{Density} = \frac{\text{mass}}{\text{volume}}$$ $$= \frac{270\ g}{540\ cm^3}$$ $$= 0.5\ g/cm^3$$	A room has a volume of 300 m³. The mass of air in the room is 3.6 kg. $$\text{Density} = \frac{\text{mass}}{\text{volume}}$$ $$= \frac{3.6\ kg}{300\ cm^3}$$ $$= 1.2\ kg/m^3$$

Material	Density (g/cm³)
air	0.001
flour	6.0
gold	19.3
ice	0.9
lead	11.7
petrol/gasoline	0.7
silver	10.5
water	1.0

⬆ Some typical densities.

Q

1 Why can you not measure the volume of water in a measuring cylinder accurately unless you look straight at the scale?

2 Look at the table of typical densities. You have a 1 cm³ volume of each of the materials.

  a List them in order of smallest mass to biggest mass.

  b Which is the odd one out? Why?

3 Calculate the densities of the following materials.

  a Material A has a volume of 2 cm³ and a mass of 30 g.

  b Material B has a volume of 10 cm³ and a mass of 8 g.

  c Is material A or material B more likely to be a liquid?

4 You are measuring the density of a liquid. You forget to subtract the mass of the empty measuring cylinder from the total mass, would your calculation of density be bigger or smaller than it should be?

!

● Density = mass/volume.

● You measure mass using a balance.

● You can measure the volume of a liquid or an irregular solid using a measuring cylinder.

Explaining density

Why are solids so dense?

Solids, liquids, and gases have very different densities. Solids are usually denser than liquids and gases are the least dense of all.

We can explain why solids are usually more dense than liquids or gases using particle theory.

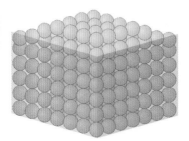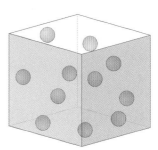

A 1 cm³ volume of a solid contains lots more particles than a 1 cm³ volume of a gas. The mass of 1 cm³ of the solid will therefore be much bigger, so it will have a much higher density.

You can make a model of solids, liquids, and gases using marbles.

- A box containing a few marbles represents the particles in a gas. To make a better model you need to shake the box so that the marbles are moving around fast because gas particles do not lie at the bottom of a container.

- A box filled with marbles is like the particles in a liquid. There is not much space between the marbles, and they are not arranged in any kind of pattern. The particles in a liquid are moving so you need to shake the box gently to show this.

- If you put the marbles in the box very carefully so they are in a regular pattern, this represents the particles in a solid. The particles in a solid are vibrating about a fixed position. It is not possible to show this using the marble model.

The table shows the densities of some different materials. They vary, but liquids and solids have roughly similar densities, while gases have much smaller densities.

Solids	Density (g/cm³)	Liquids	Density (g/cm³)	Gases	Density (g/cm³)
aluminium	2.6	water	1.00	air	0.0013
gold	19.3	petrol/gasoline	0.70	helium	0.000 18
magnesium	1.7	milk	1.03	carbon dioxide	0.0020
ice	0.9	bromine	3.12	oxygen	0.0014

Density does not just depend on the arrangement of the particles. It also depends on the *mass* of the particles. The individual particles in different materials can have very different masses.

Why do some things sink and some things float?

Floating metal buoys mark out the deep-water channel for boats coming into harbour, so that they don't hit underwater sand or rocks. Why do the channel markers float, but not the rocks at the bottom of the harbour?

You know that when a coconut floats on water, the upthrust is equal to the weight. Another way of thinking about why something floats or sinks is to compare its density with the density of the liquid. An object will float on a liquid if it has a density that is *less* than that of the liquid.

$$\text{Density of coconut} = \frac{\text{mass}}{\text{volume}}$$
$$= \frac{400\ \text{g}}{500\ \text{cm}^3}$$
$$= 0.8\ \text{g/cm}^3$$

The density of water is 1.0 g/cm³. The density of the coconut is less, so it floats.

Imagine what would happen if you had a ball with exactly the same volume as a coconut but with twice the mass. The density of the ball would be double, and it would sink.

It is not always easy to predict whether something will float or sink. Pumice is a volcanic rock that floats. Most woods float, but ironwood is a type of wood that sinks! The density of pumice is about 0.3 g/cm³ and the density of ironwood is 1.3 g/cm³.

↑ Pumice is a rock that floats.   ↑ Ironwood is a wood that sinks.

If you are swimming, it is much easier to float in salt water than in a swimming pool. This is because salt water is denser than water in the swimming pool. One of the saltiest seas is the Dead Sea between Israel and Jordan. The density of Dead Sea water is 1.24 g/cm³, compared with 1.0 g/cm³ for pure water.

Q

1 Give one reason why the density of iron is higher than the density of oxygen.

2 Explain why pumice floats and why ironwood sinks.

3 Aanjay and Faisyal are talking about why pieces of two different metals that are the same size have very different masses. Aanjay says that the particles must be different masses, and Faisyal says that it is to do with how the particles are arranged. Are they correct? What would you say to them?

4 Mercury is a liquid metal with a density of 13.6 g/cm³. Water has a density of 1.0 g/cm³.

a Gold has a density of 19.3 g/cm³. Will it float or sink in mercury? In water?

b Silver has a density of 10.5 g/cm³. Will it float or sink in mercury? In water?

c Wood has a density of 0.8 g/cm³. Will it float or sink in mercury? In water?

!

- Solids and liquids are more dense than gases because their particles are packed closer together.

- The density of a material depends on the mass of the particles and the arrangement of the particles.

- An object will float on a liquid if it is less dense than the liquid.

Questions, evidence, and explanations

Gemstones

For thousands of years people have prized jewellery. They made necklaces and rings from metals such as gold that they found in streams. They decorated the necklaces with gemstones such as red rubies and green emeralds. **Gemstones** are **minerals** found in the Earth that have distinctive colours.

Asking questions

Over 1000 years ago Al-Biruni was born in a country that is now a region of Uzbekistan. He was a Muslim scientist and mathematician who wrote books on a wide range of topics from astronomy to geography.

He was very interested in geology and minerals, and he knew that it was important to be able to identify gemstones so that people were not cheated when buying them. For example, it wasn't easy to tell the different between quartz and diamond just by looking at them. Quartz is the most common mineral on Earth, so people could make a lot of money by pretending that a lump of quartz was diamond.

⬆ Quartz and rough diamond can be hard to tell apart.

Al-Biruni wondered if it was possible to work out which stones were which by finding their densities. He knew how to measure their masses, but the stones did not have regular shapes so finding their volume was more difficult.

He designed a piece of equipment that enabled him to measure volumes **accurately**. It was a flask with a side arm similar to the apparatus shown on the left.

He filled the flask until water overflowed through the side arm into the beaker. Then he emptied the beaker and replaced it under the side arm. He then lowered the stone into the flask, and water overflowed into the beaker. He measured the volume of that water, which was equal to the volume of the stone.

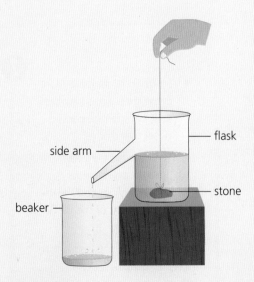

Al-Biruni used the equation for calculating density to work out the density of the gemstones.

$$\text{Density} = \frac{\text{mass}}{\text{volume}}$$

The table shows the densities that he worked out for some minerals.

Mineral	Al-Biruni's measured density (g/cm³)	Modern value for density (g/cm³)
quartz	2.58	2.58
ruby	4.01	4.4
pearl	2.7	2.7

As well as measuring density, Al-Biruni built on the work of earlier scientists by classifying gemstones according to their colour, whether white light splits up into a spectrum when it goes through the gem, their hardness, and their crystal shape. Al-Biruni was so fascinated by gemstones that he wrote a book about them.

One of the most important things about Al-Biruni's density measurements was their **precision**. It was not until over 700 years later that scientists in Europe made measurements of density with the same level of precision.

Modern science

Scientists and jewellers can still use Al-Biruni's technique to work out the density of an object to help to identify it.

Today there are also other techniques for identifying gemstones. Scientists can find the **refractive index** of a gemstone, or measure how well it conducts electricity. You learned about refractive index in Chapter 6: Light on page 121.

Al-Biruni had many research areas, building on the ideas of other scientists. He designed new techniques to make more accurate measurements. Today scientists usually specialise in one area of science, often with a team of people. They share their ideas in publications that are read all over the world and build on each other's ideas.

1 For which gemstone in the table is Al-Biruni's density the least precise?

2 For which gemstones in the table is Al-Biruni's density the most accurate?

3 Why is the precision of Al-Biruni's measurements of density so impressive?

4 Scientists work in lots of different ways.

   a Describe one way in which Al-Biruni's way of working was similar to the way scientists work today.

   b Describe one way in which it was different.

- Scientists ask questions and collect evidence.

- Creative thought is important in designing experiments to solve problems.

- Scientists share their ideas and build on each other's work.

When the Egyptians built the pyramids around 5000 years ago, there were no cranes or big machines. How did they do it? We know that they carried huge blocks of stone long distances by putting them on rollers and pulling them along, but how did they get them onto the rollers?

They used **levers**. Levers work because some forces produce a **turning effect**.

Turning forces

Forces can change the motion of objects, and they can also make things turn. Every time you close a door you are applying a **turning force** to the door. When you use a wheelbarrow or sit on a see-saw you are applying a turning force.

Levers

We can use levers to move objects that are too heavy to lift on their own. A lever is a bar that can turn when you exert a force on one end. The force that you exert on the bar is called the **effort**. The weight

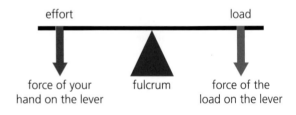

of the object that you lift is called the **load**. Sometimes the object that you are moving is also called the load. The effort produces a turning effect and the bar rotates about a **pivot** (or **fulcrum**).

You can use a lever to lift a block of stone. When you push down on one side of a lever you produce a **turning force**. This will make the other end of the lever move up, lifting the stone.

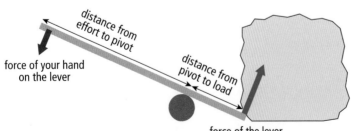

A lever works because the distance from the effort to the pivot is *bigger* than the distance from the load to the pivot. This means that you need to push down with a force that is much smaller than the weight of the stone to move it.

The lever is a **force multiplier** because it makes the effect of the force that you exert much bigger.

Levers in the world around you

Levers are not always straight bars.

Using a hammer to remove a nail is an example of a force multiplier lever. The distance from the effort to the pivot is bigger than the distance from the load to the pivot, so you need less effort to remove the nail.

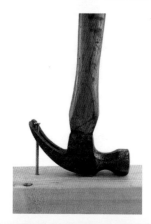

Sometimes the load is between the pivot and the effort, for example, when using a wheelbarrow. The distance from the effort to the pivot is bigger than the distance from the load to the pivot. You need to use less effort than without the lever, so you can produce a bigger load (lift a heavier weight) using a wheelbarrow than without.

In a chemistry laboratory you use tongs to hold a test tube or a beaker at a distance so you don't get burnt. In this case the effort is between the pivot and the load. This means you are using more effort than if you didn't use the tongs.

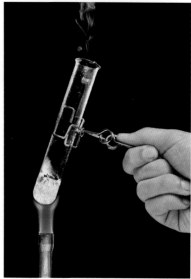

The tongs are a **distance multiplier**. If you push the handles together a small amount with your fingers, the ends of the tongs move much further.

Q

1 Copy and complete this sentence by selecting the correct words.
You need to use a **bigger/smaller** force to close a door if you push it near the hinges. This is because distance to the pivot is **bigger/smaller**.

2 The picture shows how you could make a lever to lift a stone using a plank of wood.

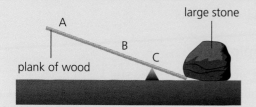

large stone

plank of wood

a Which letter (A, B, or C) is the pivot?

b Which letter (A, B, or C) shows where you should apply the *least* effort to lift the stone?

c Which letter (A, B, or C) shows where you should apply the *most* effort to lift the stone?

3 If using the tongs to hold things means that we need a bigger force, why do we use them?

- Forces can turn an object clockwise or anticlockwise about a pivot.
- A lever can be used as a force multiplier.
- Levers can produce forces that are much bigger than the effort, to lift heavy loads.

Calculating moments

On the high wire

Walking on a tightrope can be difficult and dangerous. The tightrope walker uses a long pole to help him balance. If he feels that his body is going to tilt one way, he can adjust the pole to tip him the other way. The pole cancels out the turning force to help him balance.

What is a moment?

The turning effect of a force is called a **moment**. The size of the moment depends on the force being applied and how far it is from a pivot.

Moment = force × perpendicular distance from the pivot

Force is measured in newtons (N) and distance is measured in metres (m), so a moment is measured in **newtonmetres** (Nm).

We use a turning force when we close a door.

A boy uses a force of 5 N to close a large door. The handle is 1.5 m from the hinge (the pivot). What is the moment of the force?

Moment = force × distance

= 5 N × 1.5 m

= 7.5 Nm

Balancing

When two people balance on a see-saw, we say that it is in **equilibrium**. This means that the **anticlockwise** moments and **clockwise** moments are the same. This is called the **principle of moments**:

If an object is balanced, the anticlockwise and clockwise moments are the same.

We can work out whether a see-saw will be balanced by calculating the anticlockwise and clockwise moments.

Example 1

Sunita weights 400 N. She sits 1.5 m from the pivot of a see-saw with her mother 1.0 m from the pivot on the other side. Her mother weighs 600 N. Is the see-saw balanced?

Sunita

Sunita's mother

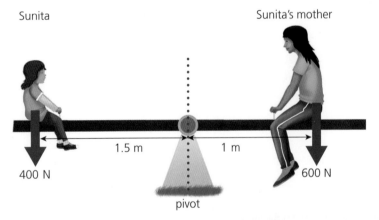

1.5 m

1 m

400 N

600 N

pivot

Anticlockwise moment = 400 N × 1.5 m	Clockwise moment = 600 N × 1.0 m
= 600 Nm	= 600 Nm

The anticlockwise moment and the clockwise moment are equal, so the see-saw is balanced.

Example 2

Davina is balancing some 10 N weights on a see-saw. Is it balanced?

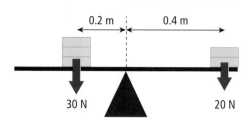

Anticlockwise moment = 30 N × 0.2 m	Clockwise moment = 20 N × 0.4 m
= 6 Nm	= 8 Nm

The anticlockwise moment and the clockwise moment are *not* equal, so the see-saw is not balanced.

Balancing a see-saw

You can use the principle of moments to work out how much weight you need to balance a see-saw or where you need to put the weight.

Problem 1: what's the weight?

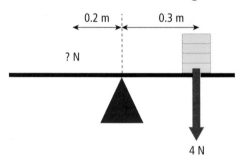

Anticlockwise moment = clockwise moment

? N × 0.2 m = 4 N × 0.3 m

? N × 0.2 m = 1.2 Nm

? N = $\dfrac{1.2 \text{ Nm}}{0.2 \text{ m}}$

Force = 6 N

Problem 2: what's the distance?

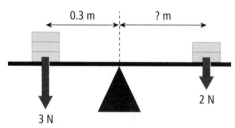

Anticlockwise moment = clockwise moment

3 N × 0.3 m = 2 N × ? m

0.9 Nm = 2 N × ? m

$\dfrac{0.9 \text{ Nm}}{2 \text{ N}}$ = ? m

Distance = 0.45 m

Q

1 Copy and complete the sentences using these words:
   small   big   moments   equilibrium
   On a see-saw, a _____ force acting a short distance from the pivot will be balanced by a _____ force acting a long distance from the pivot. The see-saw is in _____. This is the principle of _____.

2 A gymnast standing balancing on a beam will put her arms out. Why does this help?

3 In Davina's experiment above, which way will the see-saw turn?

4 **Extension:** Citra has a weight of 450 N. She sits 1.2 m from the pivot of a see-saw. How far away from the pivot must Ari sit on the other side to balance the see-saw if his weight is 600 N?

!

- The turning effect of a force is called a moment.

- A see-saw is in equilibrium if the anticlockwise moments balance the clockwise moments. This is the principle of moments.

Objectives

- How to plan an investigation
- How to reduce error and collect reliable data

Backwards and forwards

A pendulum is an object that swings backwards and forwards, like a child on a swing, or a car on a theme park ride.

Pendulum motion

The time that it takes a pendulum to swing backwards and forwards is called the **period**.

The pendulum is stationary at points A and B. It is moving fastest in the middle of the swing, at point C. The period is the time that it takes to move from A to B *and back again*.

If you wanted to measure the period of a pendulum, you would measure the time with a stopclock or stopwatch. It is very difficult to measure short lengths of time because of your **reaction time**. You learned about reaction time in Chapter 4 on pages 82–83. This is the time it takes for you to react to what you are seeing. Most people have a reaction time of about 0.2 seconds.

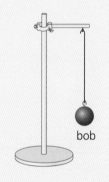

Time from centre, out one side, back through centre, out other side, back to centre is the period.

A   C   B

bob

Two approaches to the same problem

Aditi and Diya have made a pendulum from some string with a small mass on the end.

They decide to plan an investigation. They know that the first thing they need to do is to think what question they want to investigate.

I wonder what would happen if the mass was bigger?

I wonder what would happen if the string was longer?

They decide to investigate what factors affect the period of the pendulum. This is the list of variables that they think might affect the period:

- the length of the pendulum
- the mass on the end
- the amount that you pull it back.

Aditi's plan

I am going to change the length of the pendulum.

I will start with 10 cm and go up 5 cm each time until I get to 50 cm.

I will time how long the pendulum takes to swing 10 times and divide the result by 10 to get the period.

I will keep the mass of the pendulum the same and pull it back the same amount each time.

I will put my results in a table like this:

Length	Time

Diya's plan

I am going to change the length of the pendulum. I have tried out the experiment.

I am going to use a pendulum with lengths between 30 cm and 100 cm.

Each time I will time how long it takes to swing once.

I will do this three times for each length and find the average time.

I will put my results in a table like this:

Length (cm)	Time (s)	Time (s)	Time (s)	Average time (s)

Each student has identified the variable that they are going to change and the variable that they are going to measure. They have explained how they will control the other variables.

Trying things out

Diya tried out her experiment first to find out how to do the investigation. This **preliminary work** helped her to work out the range of results to take, the number of results to take, and how to get accurate results.

Aditi thought about making accurate measurements in her plan. She remembered about reaction time. She used that information to help her to decide how to measure the time for one swing. Sometimes you need to use your scientific knowledge to help you plan your investigation.

Q

1 What did Diya do to make her results more accurate?

2 What did Aditi do to make her results more accurate?

3 Describe one way in which Aditi's plan is better than Diya's plan.

4 Describe one way in which Diya's plan is better than Aditi's plan.

5 Diya found the period by measuring the time from when the mass was at the bottom of the swing. Aditi found the period by measuring the time from when the mass was at the top of the swing. Which is more likely to produce accurate results? Why?

- To make readings more accurate and reliable you repeat them and take an average.

- In an investigation you change one variable at a time. and say how you will control the other variables.

- To measure a short period you can time 10 swings and divide by 10.

Centre of mass and stability

Objectives

- State what is meant by the centre of mass (or centre of gravity)
- Explain why some objects topple over easily and others do not

Centre of mass

If you want to carry a plank of wood on your shoulder, you need to place it so that your shoulder is underneath the middle of the plank. There is a point in an object through which all of the weight of the object seems to act, called the **centre of mass** (or **centre of gravity**). The centre of mass is in the centre of the plank of wood. If this point is above your shoulder, which is the pivot, there is no turning force acting on the plank of wood.

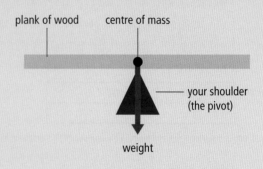

If you move the pivot closer to one end of the plank, there will be a turning force on the plank. The plank's weight acts through its centre of mass to produce a turning force that turns the plank clockwise. To balance it you will need to apply another force. This force will turn the plank anticlockwise and balance its weight.

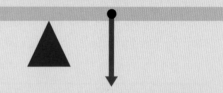

The centre of mass of a regular shape, such as a cube or a ball, is in the centre of the shape because the mass is evenly distributed throughout the object. For an irregular shape such as a broom, the centre of mass of will not be in the centre.

Why do objects topple over?

It is easier to topple some objects than others. The acrobats could fall easily, so they have to balance very carefully to make sure that they do not topple over.

We can work out why some objects are easier to topple over than others by thinking about the position of the centre of mass.

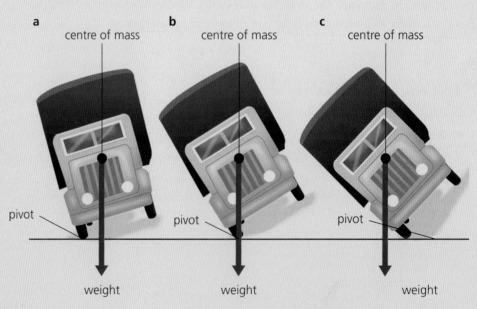

a As you start to push the toy lorry over, the centre of mass moves toward the pivot. If you stop pushing it at this point, it will fall back.

b At this point the lorry can balance because the centre of mass is over the pivot. It is hard to balance the lorry because the centre of mass has to stay exactly in line with the pivot.

c Now the centre of mass is the *other* side of the pivot, so the lorry will topple over.

centre of mass

Stability

If an object is in a position that makes it difficult to topple over, we say that it is in a **stable** position. The acrobats have to balance very carefully to make sure that they do not topple over. Their position is not very stable.

It is harder to topple objects with a low centre of mass than objects with a high centre of mass. If the centre of mass is high, then the object does not need to be tipped far before the centre of mass moves outside the pivot and it topples over. This is why high winds can blow over a tall vehicle like a truck, but are less likely to topple a car. The car has a lower centre of mass and so is more stable.

Moving your centre of mass

You can demonstrate how your centre of mass can move. If you bend over to pick up a chair, you naturally lean back slightly so that your centre of mass stays over your feet. If you stand against a wall and try and pick up the chair, you will find it impossible. You cannot lean forward and still keep your centre of mass over your feet. You feel you are going to topple over.

⬇ In July 2012 the world record for human mattress toppling was set by 1001 volunteers in Shanghai, China. The gaps between the mattresses were fixed so that when one toppled over it knocked the next one over. Each person falls safely because the mattresses are soft.

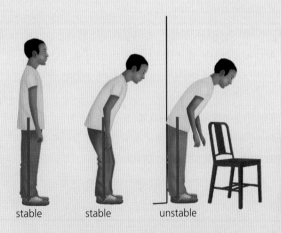

stable   stable   unstable

Q

1 If you lean back in a chair, the two back legs act as a pivot. You can only lean so far back without falling over. Explain why in terms of your centre of mass and a turning force.

2 Explain why it would become more and more difficult for the acrobats to keep their balance as more acrobats are added to the top.

3 Sketch a diagram of a brush with a long handle that you use to sweep the floor. Label where you think the centre of mass might be. Explain your answer.

4 It is almost impossible to balance a pencil on its point. Why?

!

- The centre of mass (or centre of gravity) of an object is the point through which the mass of the object seems to act.

- Objects with a high centre of mass are easier to topple than those with a low centre of mass.

1 Copy and complete the table to give the units for each quantity. The first one has been done for you. You may need to write more than one answer in each box.   [7]

Quantity	Unit name	Unit letter
time	second minute hour	s m h
force		
area		
pressure		
density		
volume		
mass		
moment		

2 Copy the table. Use the equations for pressure to complete it: [5]

$$\text{Pressure} = \frac{\text{force}}{\text{area}}$$

$$\text{Force} = \text{pressure} \times \text{area}$$

$$\text{Area} = \frac{\text{force}}{\text{pressure}}$$

Force (N)	Area (m²)	Pressure (N/m²)
100	2	
250		5
	6	200
5	0.1	
0.01		0.04

3 Two glasses are filled with liquid to the same height. Glass W is filled with water and glass F is filled with fruit juice. Fruit juice has a higher density than water.

a Will the pressure in the glass of fruit juice be bigger or smaller than the pressure in the water? Explain your answer. (*Hint:* think about why pressure changes with depth.)   [2]

b Write the letters to show these pressures in order from the lowest pressure to the highest pressure.

   A the pressure at the bottom of glass F   [1]

   B the pressure at the bottom of glass W   [1]

   c the pressure half-way up glass W   [1]

4 A mechanic is using a hydraulic jack to lift a car. Here is a simple diagram of the hydraulic jack.

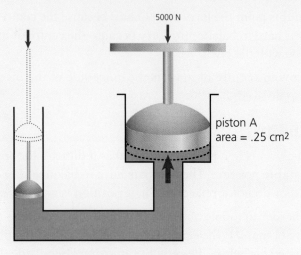

The weight of the car is 5000 N. The area of the piston that lifts the car is 0.25 m².

a What pressure is needed in the liquid to lift the car?   [2]

The area of the piston that the mechanic pushes down is 0.01 m².

b What force does the mechanic need to use to lift the car?   [2]

c Why are liquids rather than gases used in hydraulic machines like this?   [1]

5 Anyam blows up a balloon and seals it.

a If Anyam leave the balloons in the sunshine so that the gas inside gets hotter, what do you think will happen?   [1]

b Explain your answer to **a**.   [1]

6 A student wants to find the density of a brick. He measures the sides of the brick.

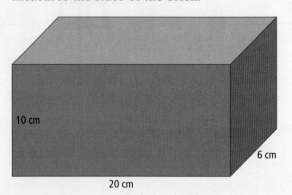

a What is the volume of the brick?   [2]

b He measures the mass of the brick and finds that it is 0.24 kg or 240 g. What is the density of the brick?   [2]

7 Here are the densities of some materials.

Material	Density (g/cm³)
air	0.001
ice	0.9
iron	7.9
plastic	1.4
water	1.0
wood	0.7

For each of these statements, write true or false.

a 10 cm³ of iron has a smaller mass than
   10 cm³ of wood.   [1]

b 2 cm³ of ice has a bigger mass than 1 cm³
   of water.   [1]

c 1 cm³ of plastic has a bigger mass than
   100 cm³ of air.   [1]

d 5 cm³ of plastic has a smaller mass than
   1 cm³ of iron.   [1]

8 A fish can move higher and lower in the water.
   It has a sac inside it called a swim bladder.
   If the fish fills the sac with air from its gills,
   it will float up to the surface. Explain why.   [2]

swim bladder allows
fish to float or sink

9 Which of these statements about the way that
   scientists work today is *not* correct?   [1]

a Scientists share their ideas and scientific
   evidence in publications that others read.

b Scientist use creative ideas to interpret the
   evidence that they collect in experiments.

c It is unusual for scientists to read about the
   work of other scientists.

d It is possible to make measurements that
   are more precise and accurate using new
   technology.

10 Chinonye is timing his friend on a swing.
   He measures the time it takes for 10 swings.

a The time for 10 swings is 6 seconds.
   What is the period of the swing?   [1]

b Why is it more accurate to measure the time
   for several swings and divide by the number
   of swings?   [1]

Chinonye measures the time for different children
to complete 5 swings. Here are his results:

Mass (kg)	Period (s)
350	1.2
400	1.3
450	1.1
600	1.2

c What conclusion can you make about the
   link between mass and period?   [1]

11 You are *planning* an investigation. Which of
   these things do you *not* do?   [1]

a decide on a question to investigate

b take precise and accurate results

c do preliminary work

d decide which results to record

e work out how to work safely

12 Calculate the moment of each of these forces.

a a force of 10 N acting 2 m from a pivot   [1]

b a force of 2 N acting 0.4 m from a pivot   [1]

c a force of 0.1 N acting 0.2 m from a pivot   [1]

13 Jamal sits at the end of the see-saw. Jamal has a
   weight of 200 N.

a Maria, weight 250 N, sits on the other end, an
   equal distance from the pivot. What will happen
   to the see-saw?   [1]

b Jamal's brother has a mass of 400 N.
   He tries to balance the see-saw instead
   of Maria. Where should he sit on the
   see-saw to make it balance?   [2]

c Ryan sits on one end of a see-saw. He has
   a mass of 450 N. Can Jamal sit on the other side
   to make the see-saw balance? Explain
   your answer.   [2]

Electrostatic phenomena

Strange happenings

If you rub a balloon against your clothes it will stick to the wall. This balloon will then deflect a stream of water from a tap, or pick up small pieces of paper. Why?

These are examples of **electrostatic** phenomena. A phenomenon is something that you can observe or sense. These things happen because when you rub the balloon you **charge** it.

Charging and discharging

Kunala rubs a polythene rod with a cloth. He hangs it up by a piece of string.

He rubs another polythene rod and brings it close to the first rod. The

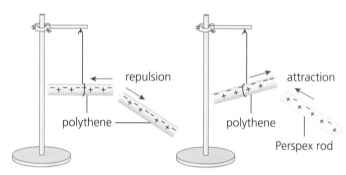

first rod is **repelled** by the second rod. Then he rubs a Perspex (clear plastic) rod and brings it close to the polythene rod. This time the polythene rod is **attracted** to the Perspex rod.

The rods are repelled or attracted because they have become electrically charged.

Why do objects become charged?

There are two types of electric charge, **positive (+) charge** and **negative (−) charge**.

- A positive charge will repel another positive charge.

- A negative charge will repel another negative charge.

- Positive charges attract negative charges.

We say that *like charges repel, unlike charges attract.*

When Kunala rubs the polythene rod it becomes negatively charged. When he brings another charged polythene rod close to it, it is repelled because the charge on both rods is the *same*.

When he brings a Perspex rod near the polythene rod, it is attracted. This is because the charge on the Perspex rod is positive – the rods have unlike charges and so they attract.

Where does the charge come from?

All materials are made of tiny particles called atoms. Every atom contains three types of particle: protons, neutrons, and electrons.

- Protons are **positively charged**.
- Neutrons have no charge.
- Electrons are **negatively charged**.

The diagram shows a simple model of an atom. There are protons and neutrons in the centre of the atom, called the **nucleus**. The electrons orbit the nucleus.

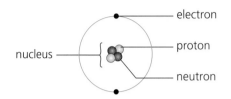

In an atom there is always the same number of electrons and protons, so an atom has no charge overall. We say that it is **neutral**. There is no *net* charge. If you remove one or more electrons from the atom it becomes positively charged, because there is now more positive than negative charge.

When you rub a polythene rod, electrons are transferred from the cloth to the rod. Electrons have a negative charge so the rod has a net negative charge. The cloth has a net positive charge. It does not have enough electrons to **neutralise** the positive charge.

When you rub a Perspex rod electrons are transferred from the rod to the cloth. The Perspex rod has a net positive charge because it does not have enough electrons to neutralise all the protons in its atoms. It is important to realise that you are not transferring *all* the electrons, just some of them.

You always have the same total amount of charge in the rod and the cloth. Charge cannot be created or destroyed, just like energy.

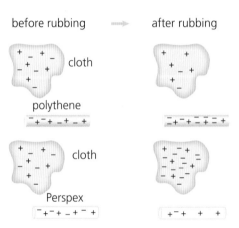

Conductors and insulators

Static means stationary or standing still. We see electrostatic phenomena, such as the rods repelling and attracting each other, because the polythene and Perspex are insulators. In an **insulator** the electrons do not move around. They stay on the rod once you have rubbed it.

If you rub a metal rod with a cloth it will also become charged, but the metal is a **conductor**. Electrons can move in it. Any extra electrons will move through the metal to your hand, through you to earth.

Q

1 Explain why an atom is neutral.

2 A teacher rubs a Perspex rod with a cloth. It becomes charged.

   a What is the net charge on the cloth?

   b What is the net charge on the rod?

   c Would the rod attract or repel a charged polythene rod? Explain your answer.

3 Why can plastic become charged if you rub it, but not metal?

4 A student says that the rods repel because they have poles like magnets. Is she correct? Explain your answer.

!

- There are two types of charge: positive and negative.

- An object that has equal numbers of positive and negative charges is neutral.

- If electrons are transferred from one object to another, each object will become charged.

- In a conductor the electrons are free to move, but in an insulator they are not.

9.2 Dangers of electrostatic phenomena

Objectives

- Describe how electrostatics phenomena can dangerous

- Explain how the risk of damage from electrostatic phenomena can be reduced

There are lightning storms all over the world every day. Lightning is an example of an electrostatic phenomenon.

Sparks and current

If enough charge builds up due to friction, then a **spark** will form. A spark is charge moving through the air, heating up the air. We see light and hear a sound. The moving charge is an **electric current**.

Air does not usually allow a current to flow through it. It takes a lot of charge to build up for the air to **conduct** electricity.

This is what happens in a thundercloud. Inside the cloud the air moves around and forms a region that is positively charged and a region that is negatively charged, just like rubbing a rod with a cloth. Negative charges at the bottom of the thundercloud can make a huge spark as they travel to Earth. We see that spark as lightning.

Sometimes there is a movement of charge between the positive and negative regions inside the cloud, and the lightning lights up the cloud.

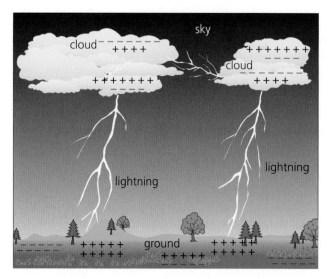

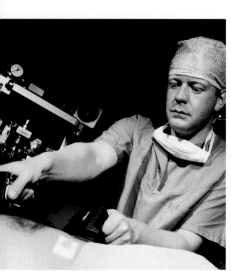

Electric shocks

Lightning can be very dangerous. If it strikes a person it can break their bones, cause burns, or even kill them. The charge travels through the person to earth. The electric current can interfere with their heart or even stop their heart beating.

Doctors in hospitals can use an electric current to try to restart a patient's heart if it stops beating. This can save their life.

Reducing risk

In a thunderstorm trees are likely to be struck by lightning, so you should never shelter under a tree. Buildings can be damaged if lightning strikes them – a high tower might be struck and fall down.

Risk is a combination of the probability of something happening and the consequence if it did. We cannot change the probability of lightning hitting a building, but we can change what happens if it does.

Lots of high buildings have a **lightning conductor** attached to them. A lightning conductor is a thick strip of metal such as copper running down the wall. The strip is attached to a copper plate buried underground. When lightning strikes the building electrons flow down the strip to the plate, and into the ground. The metal conducts electricity much better than the building. This saves the building from being damaged.

If you want to reduce the risk of damage by an electric current, you can **earth** the object. This means connecting it to the Earth with a conductor, just like to the lightning conductor.

Fuel in cars and planes

When cars or planes are refuelled there is a risk of sparks. This would be dangerous because the fuel could catch fire. The friction of the fuel moving down the pipe can cause a charge to build up.

Fuel pipes are earthed by attaching a metal wire to the pipe. Any charge that builds up will flow down the wire to the Earth. This reduces the risk of a spark, and an explosion.

Electronics

Engineers who make circuits for TVs or phones need to reduce the risk of damage to the delicate electrical components. They often wear special wristbands connected to a metal wire that goes to the ground. Any build-up of charge flows down the wire to earth rather than through the delicate components.

Everyday electrostatics

Sometimes people feel a small shock when they touch a car door. This is because the car has become charged by friction as it has been moving along. You might sometimes hear the crackle of sparks as you remove a piece of clothing, or feel a small shock when you touch a door handle.

Q

1 What is lightning?

2 Explain what is meant by 'earthing'.

3 Explain how the engineer's wristband can protect the components.

4 A student notices that she gets a shock when she touches a metal door handle. She has been charged by walking across a carpet.

   a Would she get a shock if the handle was made of plastic?

   b Explain your answer.

- Sparks happen when the air conducts electricity.

- Lightning is a big spark that can cause damage to people and buildings.

- Lightning conductors reduce the risk of damage to buildings.

- Earthing reduces the risk of sparks and shocks.

Digital sensors

Many mobile phones and computers use touch screen technology. You can touch a keypad on the screen to dial a number, or take a photograph with a digital camera. How do touch screens and digital cameras work?

Touch screens

Touch screens work using stored electric charge. One of the simplest ways to store charge is using a **capacitor**.

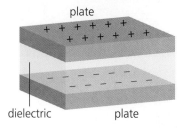

plate

dielectric   plate

A capacitor consists of two metal plates separated by an insulator called a **dielectric**. A battery is connected to both plates. It pulls the electrons off one plate and pushes them onto the other. When the battery is disconnected the electrons will stay there because they cannot move through the dielectric.

A touch screen works like a capacitor. One plate of the capacitor is in the screen and the other plate is your finger. A sensing circuit connected to the touch screen can work out where your finger is. If you are dialling a phone number the circuit can detect which number you have selected. Some screens have sensors that can work out where your finger is *and* how it is moving. Your finger can make a picture bigger or smaller, or move it around the screen.

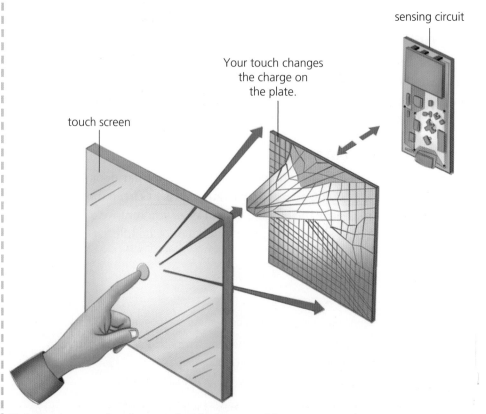

sensing circuit

Your touch changes the charge on the plate.

touch screen

Touch screen technology is developing quickly. Some touch screens don't work when you are wearing gloves, but there are types that do. There are some touch screens that don't use capacitors – they work in a different way.

Digital cameras

At the back of a digital camera is a **charge-coupled device** (**CCD**). This is a grid of components that work like capacitors.

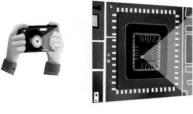

When radiation such as light hits each of the components in the CCD, charge is produced. The charge is stored in the component's capacitor. The amount of charge stored depends on the intensity of the light. The charge on each component is transferred to a circuit that produces a type of **signal** called a **digital signal**. The camera converts this signal into an image. You may have noticed that there is a slight delay when you are taking a picture with a digital camera or a phone. This is because the charge is being moved from the CCD.

Any image made by a digital camera is made up of small dots called picture elements, which are often called **pixels** for short (from <u>pic</u>ture <u>el</u>ement).

Colour pictures need to mix red, green, and blue light to make a colour image. As you saw on page 126 all the different colours of light can be made using the three primary colours of light.

Electric fields

There is a gravitational field around the Earth. Any mass placed in that field will experience a force. There is also a magnetic field around a bar magnet. Iron filings placed in that field will experience a force.

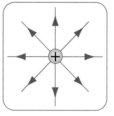
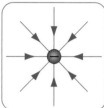

There is a field around a charge. It is called an **electric field**. You can draw electric field lines in the same way as magnetic field lines. You might recognise some of the patterns.

When you touch a touch screen you are changing the electric field between your finger and the plate inside the screen. The phone detects these changes and uses them to work out where your finger is and how it is moving.

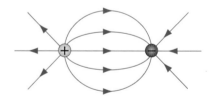

1 What would happen in a capacitor if the dielectric was not made out of an insulator?

2 In what way is a touch screen similar to a CCD? In what way is it different?

3 Describe what is meant by a pixel.

4 Look at the diagrams of the fields around charges.

  a Draw a diagram of the electric field around two positive charges.

  b Where have you seen a similar field diagram before?

- Touch screens work using electric charge.

- Capacitors can store electric charge.

- Digital images are made up of picture elements called pixels.

- A digital camera contains a grid that produces charge when light hits it.

Electric circuits

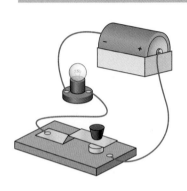

Circuit components

Televisions, movie projectors, loudspeakers, and many other devices that we use for entertainment run on electricity. Hospitals use machines to keep people alive that run on electricity. They contain **electric circuits** that use electrical **components**.

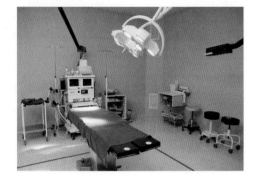

You can make a very simple circuit using a lamp, a battery, and some wire. This is the circuit that you find in a torch. If you add a switch to the circuit you can turn the lamp on and off just like you can turn a torch on and off.

Circuit symbols

Circuits can get quite complicated, and it takes too long to draw pictures of the components in a circuit. We can use **circuit symbols** for each of the components instead of drawings. Circuit symbols show which components are in a circuit.

A **cell** is something that many people refer to as a battery. In physics a **battery** contains two or more cells.

A cell has a positive and a negative **terminal**. If you connect cells together in a battery, you must make sure that you do not connect two positive terminals together (or two negative terminals together) or it will not work. On the circuit symbol the long line represents the positive terminal, and the short line represents the negative terminal.

Component	Symbol
wire	————
cell	⊣⊢
battery of cells	⊣⊢-⊣⊢
lamp	⊗
open switch	─o o─
closed switch	─o─o─
buzzer	⊔
motor	─(M)─

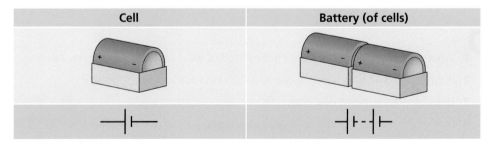

Cell	Battery (of cells)

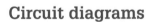

Circuit diagrams

You can join circuit symbols in a **circuit diagram** to show how the components in your circuit are connected.

This is the circuit diagram for a torch. The diagram shows that the switch is open.

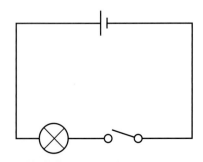

The lamp is not lit. The switch is a break in the circuit and you need a complete circuit for the lamp to work. If you press the switch the circuit will be complete and the lamp will light up.

Conductors and insulators

You can use a circuit like the lamp circuit to find out whether objects conduct electricity or not. An object or a material that conducts electricity is a conductor, and an object or a material that does not is called an insulator.

Mawar and Harta are investigating conductors and insulators. They set up a circuit to work out which things are conductors and which are insulators.

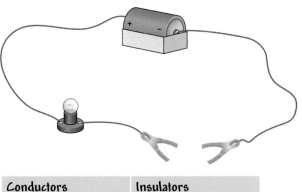

I will put different materials in the gap. If the bulb lights up that means the material is a conductor.

All the conductors are made of metal.

Conductors	Insulators
metal spoon	paper
coin	plastic spoon
nail	wood
aluminium foil	string

Some people think that *only* metals conduct electricity but that is not correct. Carbon is the only non-metal that conducts electricity. The material in the centre of a pencil is graphite, which is a form of carbon. The graphite will conduct electricity but the wood around it will not.

The wires that connect electrical appliances to plugs are made of copper, which is a metal. It would be very dangerous to have bare wires because you could get an electric shock if you touched it. The copper wire is covered in plastic, which is an insulator.

Q

1 Look at the electric circuits below.

A   B   C

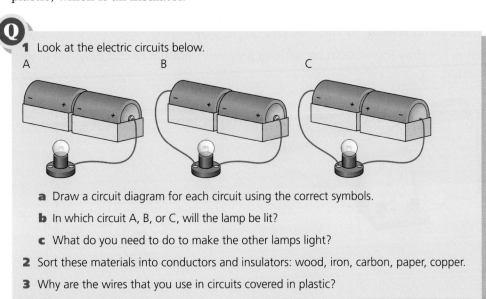

  a Draw a circuit diagram for each circuit using the correct symbols.

  b In which circuit A, B, or C, will the lamp be lit?

  c What do you need to do to make the other lamps light?

2 Sort these materials into conductors and insulators: wood, iron, carbon, paper, copper.

3 Why are the wires that you use in circuits covered in plastic?

!

- We use circuit symbols for components to draw circuit diagrams.
- A complete circuit is needed for components in the circuit to work.
- Conductors allow electricity to flow through them, but insulators do not.

Current: what is it and how we can measure it

Objective

- Describe what an electric current is and how we measure it
- Describe what is meant by a series circuit

What is electric current?

When people talk about 'electricity' they are usually talking about electric current.

An electric current is the flow of **charge**. In a circuit made of metal wires the flowing charge is caused by the movement of **electrons**. Electrons are charged particles.

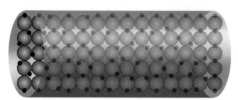

Inside a copper wire there are **atoms** of copper. These atoms are arranged in a regular pattern. Some of the electrons on the outside of the copper atoms are not strongly bound to the atom and can move around. You might hear people talk about a 'sea of electrons' inside the metal wire.

When you connect a battery to a circuit this provides a push to make the electrons move. The moving electrons make an electric current. The **current** is the amount of charge flowing per second.

How do we measure current?

Current is measured in **amperes (A)** or **amps** for short. Small currents are measured in **milliamps (mA)**. 1 mA = 0.001 A, or one-thousandth of an amp.

You measure current with a meter called an **ammeter**. The circuit symbol for an ammeter is:

There are two types of ammeter. One type shows the current using a needle on a scale. The other type shows the current on a digital scale. You can find out more about how to use these meters on page 240–1.

Here is a drawing of a circuit with an ammeter, and the circuit diagram to go with it.

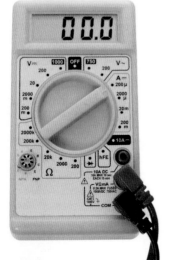

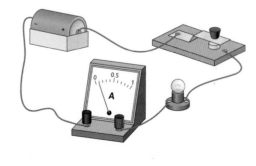

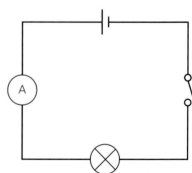

The ammeter is measuring the charge that is flowing past it per second. You can think of an ammeter as being like a person with a stopwatch counting how much charge goes past each second. If more charge flows in the wire per second, the current will be bigger.

The ammeter does not affect the current. If you connected up the circuit with no ammeter the brightness of the lamp would be the same.

Series circuits

A circuit with a single loop is called a **series** circuit. The current will only flow if it is a complete circuit. You can turn the lamp on and off with a switch, because an open switch is a break in the circuit.

In a series circuit with several lamps, you cannot turn the lamps on and off separately. They are either all on or all off. If one lamp breaks, or 'blows', then all the lamps go out.

You can use an ammeter to measure the current at different points in a series circuit.

The current at P, Q, and R is the same, so the reading on each of the ammeters is the same. Current is not used up in an electric circuit. It is the same in every part of the circuit.

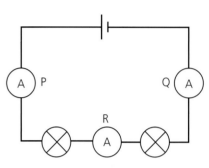

conventional current

Conventional current and electron flow

In a circuit with a cell and a lamp, a current will flow. We normally draw the current flowing from the positive terminal of the battery to the negative terminal of the battery. This is **conventional current**. This is because scientists have agreed that current flows from positive to negative.

But the charge on an electron is negative. That means that the electrons actually flow from negative to positive. They travel in the opposite direction.

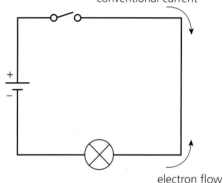

electron flow

Q

1 A student says that current is the amount of charge. Is he correct? Explain your answer.

2 Copy and complete these sentences.

   a An ammeter measures the _____ flowing per_____.

   b A _____ current means that more _____ flows per second.

   c If a bulb breaks in a series circuit the current will be _____.

3 Why do you only need one switch in a series circuit?

4 A student connects up an ammeter in a circuit with a lamp and a cell. The reading is 0.5 A. She moves the ammeter to the other side of the lamp.

   a What will the ammeter read now?

   b Explain your answer.

5 Some people think that there will be more current in the circuit near the positive end of the cell than the negative end. Is that correct? Explain your answer.

!

- Current is the flow of charge (electrons) per second.

- Current is measured in amperes (A) with an ammeter.

- The current in a series circuit is the same everywhere.

Parallel circuits

Series and parallel circuits

Neela is making some circuits using lamps, a battery, and switches. She makes a series circuit with a battery, one lamp, and a switch. Then she adds another loop containing a lamp and a switch that uses the same battery as in the first circuit. Here is a picture of the circuit that she made.

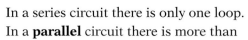

In a series circuit there is only one loop. In a **parallel** circuit there is more than one loop. The cell or battery pushes the current around each loop.

Neela made a parallel circuit when she made the circuit with two loops. She could draw a picture of her circuit like this.

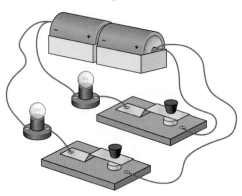

In physics we usually draw parallel circuits like this.

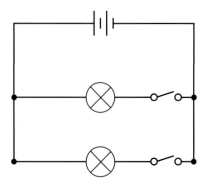

The three drawings above show the *same* circuit. In all of the circuits each bulb is connected directly to the battery. It works independently.

The different loops of the circuit are sometimes called branches. Parallel circuits are sometimes called branching circuits.

In a series circuit, if one lamp breaks then all the lamps go off. In a parallel circuit this is not the case. If one lamp breaks then the other lamps will still work.

You can put switches into each branch of the circuit to control each lamp. Each lamp can be turned on and off independently.

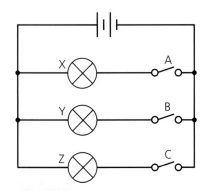

You can control the lamps X, Y, and Z separately, using switches A, B, and C.

Current in parallel circuits

Neela decided to measure the current in the branches of a parallel circuit.
She connected up a circuit with a battery, lamps, ammeters, and switches.

This is what she wrote in her book.

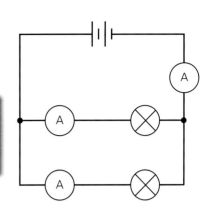

I connected up my circuit and measured the current in each of the branches and next to the battery. The current in each branch was 0.4 A. The current next to the battery was 0.8 A.

She replaced the lamps with buzzers and repeated the experiment.

With the buzzers the current in each branch was 0.1 A and the current next to the battery was 0.2 A.

Finally she put a lamp in one of the branches and a buzzer in the other branch.

This time the current in the branch with the buzzer was 0.1 A, in the branch with the lamp it was 0.4 A, and next to the battery it was 0.5 A.

Finding the pattern

Sometimes you can see patterns in your results more clearly if you write them in a table.

Now Neela can see a clear pattern in her results. The current in the wire next to the battery is equal to the sum of the currents in the branches. She writes down a rule.

Components	Current in branch 1 (A)	Current in branch 2 (A)	Current next to battery (A)
branch 1: lamp branch 2: lamp	0.4	0.4	0.8
branch 1: buzzer branch 2: buzzer	0.1	0.1	0.2
branch 1: lamp branch 2: buzzer	0.1	0.4	0.5

Current in branch 1 + current in branch 2 = current next to the battery

Models for electric circuits

Using models

You cannot see a current flowing in a wire. This can make it difficult to understand what is happening in an electric circuit. To make it easier to work out what is happening, and to predict what might happen in different circuits, scientists use different models.

Whenever you use a **model** to try to explain something in science you need to remember that all models have limits. That means that there may be things that they do not explain very clearly, or at all.

Y   X

The rope model

One way of modelling what is happening in a circuit is to use a piece of rope. The rope is tied to make a continuous loop. One person is the 'battery' and another person is the 'lamp'.

X moves the rope around the circle by pulling it through his hands. This is like the battery pushing the charges around the circuit.

Y supports the rope in his hands and can feel it moving through. His hands get warm. This is like what happens in a lamp. The filament of a lamp gets so hot that it gives out light.

The rope	…is like …	the charges (electrons) in the wire.
Person X pulling the rope	…is like …	the cell or battery pushing the charge
The amount of rope going through his hand per second	…is like …	the current (the charge flowing per second).
Person Y gripping the rope	…is like …	a component, like a bulb, in the circuit.

The rope model is a good model because it helps us to visualise what is happening. It also helps us to understand what is really happening in an electric circuit.

The rope model shows…	So in an electric circuit…
Person X does not produce the rope.	The charges are not produced by the battery. They are already in the wires and components.
The same amount of rope passes through any point every second.	The current is the same everywhere in a series circuit.
All parts of the rope start to move as soon as person X starts passing it through.	The current starts to flow as soon as the battery is connected.
The rope transfers energy from person X to person Y.	The current transfers energy from the battery to the components in the circuit.

The factory or people model

Another model for an electric circuit uses the idea of a factory that makes things that are sold in a shop.

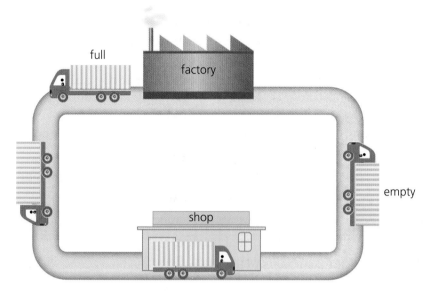

In this model the factory represents the battery, the trucks are like the charges, and the shop is like the lamp.

There are lots of other ways that you can make a model like this. You could use a person holding a pot of sweets instead of the factory, people carrying the sweets instead of trucks, and another person eating the sweets as the shop. The people walking around the circuit collect sweets and give them to the person eating them (the lamp). They walk back to the pot of sweets again. They represent the charges flowing in the circuit. The sweets represent the energy that is being transferred from the battery to the lamp.

The water circuit model

In this model the pipes are like the wires in a circuit. The water flows in the wires because the pump pushes the water around the pipe. You can stop the flow of water using a tap just like a switch can stop the current in a circuit. The meter measures the flow of water just like an ammeter in a circuit.

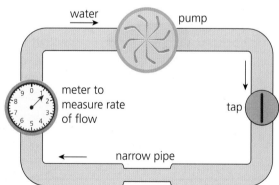

Q

1 In each of the three models for an electric circuit, what represents a flat battery?

2 How would you use the rope model to:

  a make a parallel circuit

  b explain why the currents in a parallel circuit add up?

3 In the sweet model what would represent an ammeter?

4 In the water model what does the water represent?

5 A student is wondering how to put a switch in the truck model and the rope model. What would you say to her?

!

- Models helps us to work out what is happening in a circuit.

- Models can be used to predict what might happen if you change something in a circuit.

- All models have limitations.

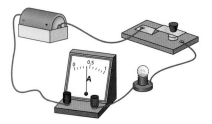

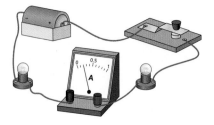

Series circuits

You can change the current in a series circuit by changing the number of components, or the type of components. You can also change the current by changing the number of cells.

Changing the number or type of component

Nanda connects up a circuit with one cell, a switch, one lamp, and an ammeter. When he presses the switch the lamp comes on, and is lit to **normal brightness**.

Then he adds another identical lamp. Two things happen.

- The lamps are now dimmer than normal brightness.
- The reading on the ammeter is *half* what it was with one lamp.

Each component such as a lamp provides a **resistance** to the flow of charge. This is like person Y holding the rope in the rope model. If two people held the rope instead of one, this would slow the rope's movement down more. Two lamps provide twice the resistance of one lamp, so only half the amount of charge flows per second. The current is halved.

Next Nanda connects up another circuit with one cell, a switch, and an ammeter, but this time he adds a buzzer instead of the lamp. He presses the switch and the buzzer makes a noise. He notices that the current is less than it was in the circuit with one lamp of normal brightness.

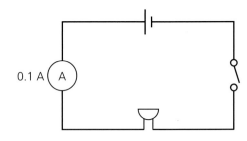

I wonder why the current is different? There is only one component in the circuit.

Different components provide different resistances to the flow of charge. If the current is less with the buzzer, that means that the buzzer provides a *bigger* resistance to the flow of charge than the lamp. This is like someone squeezing the rope a bit harder.

Changing the number of cells

Nanda starts with his first circuit containing one cell, a switch, one lamp, and an ammeter. This time he adds a second cell. Now the lamp is brighter than normal, and the reading on the ammeter is *twice* what it was before.

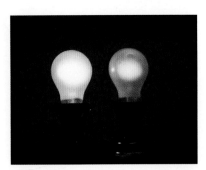

Adding cells will *increase* the current in a series circuit. Adding a second cell is like person X pulling the rope through twice as hard. Twice as much rope will pass a point each second. Person Y holding the rope will feel his hands getting warmer. This is like the lamp giving out more light.

Parallel circuits

Nanda wonders what happens when you change the components in a parallel circuit. He connects up a circuit with one cell, a switch, one lamp, and an ammeter. The ammeter reads 0.4 A and the lamp has normal brightness.

Next he adds another lamp in parallel with the first lamp. Both of the lamps are normal brightness, but the ammeter now reads 0.8 A. Adding more lamps in parallel *increases* the total current.

> I understand.
> The current in each lamp is 0.4 A, so the total current is 0.8 A.

To check that he is right, Nanda puts an ammeter in each branch of the circuit by the lamps. They both read 0.4 A. One way to think about what is happening is to say that the current splits when it gets to junction P and joins up again when it gets to junction Q.

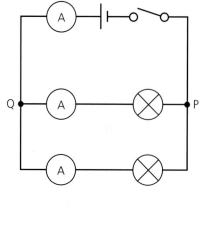

You can think about this using the rope model. Each of the lamps is in a separate loop. This picture shows how that might work. The green dotted line shows one rope loop, and the blue dotted line shows another rope loop.

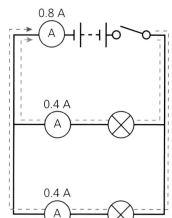

The current in a circuit depends on the voltage and the resistance.

$$\text{Current (A)} = \frac{\text{voltage (V)}}{\text{resistance (ohms)}}$$

Resistance is measured in ohms, which has the symbol Ω.

This means that you can calculate the resistance of a component using the current and the voltage.

$$\text{Resistance (ohms)} = \frac{\text{voltage (V)}}{\text{current (A)}}$$

---

1 a In which circuit below A, B, or C, will the current be biggest?

   b In which circuit, A, B or C, will the lamps be dimmest?

2 Circuits X and Y are series circuits that each contain one lamp. One of the circuits has more cells in the battery than the other. In circuit X twice as much charge flows through a lamp per second than it does in circuit Y.

   a In which circuit is the current bigger?

   b Which circuit has more cells in the battery?

   c There are 2 cells in circuit A. How many are there in circuit B?

3 What represents resistance in the water model of a circuit?

4 a The current near the battery in a parallel circuit with two lamps is double the current in a series circuit with one lamp. How does the rope model explain this?

   b Explain how having two loops in a circuit is the same as saying that the current splits at a junction.

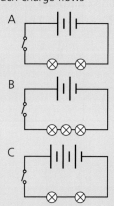

- In a series circuit the current decreases if you add more components.

- Different components have different resistances.

- In a parallel circuit the current increases if you add more branches to the circuit.

9.9 Voltage

Objective

- Explain what is meant by voltage. Describe how to measure voltage in series and parallel circuits

Different voltages

Circuits don't work without a battery or cell. Batteries (cells) come in different shapes and sizes, and they also have different voltages. Batteries are a store of chemical energy.

What is voltage?

A force is needed to make the charges in a circuit move. This force is provided by the battery. The size of the push depends on the **voltage**. The voltage also tells us about the energy that is transferred to the components by the current.

If the voltage is bigger, the battery will give a bigger push to the charges in the wires. That means that more charge will pass a point in each second, so the current will be bigger.

If the voltage is bigger, more energy is transferred to the components in the circuit. That means that lamps will be brighter and buzzers will be louder.

Measuring voltage

Voltage is measured in **volts** (V) using a **voltmeter**. The circuit symbol for a voltmeter is:

You use a voltmeter to measure the voltage *across* a component. Voltmeters are connected in parallel, not in series like ammeters.

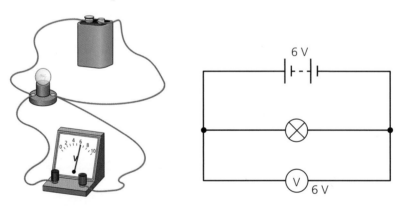

The chemical energy stored in the battery is transferred to the components by the current. The reading on the voltmeter tells you how much electrical energy is transferred to the component by the charge flowing through it.

You can connect a voltmeter across the terminals of a battery or cell to find out how much energy it can give to the charges flowing round a circuit. The voltage of this cell is 1.5 V.

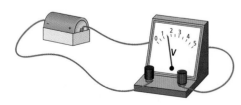

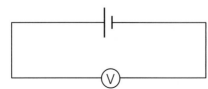

Voltage in series circuits

The voltage of the battery is 6 V. Each lamp changes energy stored in the cell to light energy and thermal energy. Half the energy from the cell is transferred in each lamp, so the reading on each voltmeter is 3 V.

You can use the factory model to help you to work out what is happening. A factory supplies two shops. Each lorry loads 6 boxes at the factory. This is like the 6 V of the battery. The lorry delivers 3 boxes to each shop. The lorry has 3 fewer boxes after delivering to each shop. Each lamp has a voltage of 3 V.

If the components in the circuit are not the same then the voltages across them will be different. If you connect a circuit with a lamp and a motor, the voltage across the motor will be bigger than across the lamp. The voltage across the lamp and the voltage across the motor will still add up to the voltage across the battery.

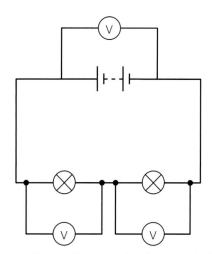

↑ The reading on each voltmeter is half the reading across the battery if the two components are the same.

Voltage in parallel circuits

In a parallel circuit the voltage across each lamp is the same. It is also the same as the voltage across the battery. All three voltmeters will read the same. This is because each component in a parallel circuit is independent of the other components. We could add more lamps in parallel and the reading on the voltmeters would still be the same.

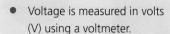

↑ The reading on all three voltmeters is the same.

1   How many 1.5 V cells would you need to provide a voltage of 9 V?

2   Explain why voltmeters are connected in parallel but ammeters are connected in series.

3   Why do the voltages across components in a series circuit add up to the voltage across the battery?

4   Dilip connects up a series circuit with a lamp and a buzzer. The voltage of the battery is 6 V.

   a   If the voltage across the lamp is 2 V, what is the voltage across the buzzer?

   b   Will the current in the lamp be bigger, smaller, or the same as the current in the buzzer? Explain your answer.

5   A student says that the circuit drains the battery because the charge is used up. What would you say to him?

- Voltage is measured in volts (V) using a voltmeter.

- The voltages across components in a series circuit add up to the voltage of the battery.

- The voltage across each branch of a parallel circuit is the same, and is equal to the voltage of the battery.

Selecting ideas to test circuits

Chetana, Dipali, and Lakshima have been learning about current and voltage. Their teacher explained that the current flowing in a component depends on the voltage of the battery and the resistance of the component. If the voltage stays the same, then increasing the resistance will make the current smaller.

Changing resistance

Asking questions

Chetana has noticed that the filament of the lamps that they use in their experiments is made of very thin wire.

Dipali looks at the filament. She notices that it is curled up and thinks that it must be very long.

Lakshima wonders what the filament is made of.

Objectives

- Select ideas that can be investigated
- Plan an investigation

⬆ Chetana

I wonder how the resistance of the wire depends on the thickness of the wire.

⬆ Dipali

I wonder how the resistance of the wire depends on the length of the wire.

⬆ Lakshima

I wonder how the resistance of the wire depends on the material that the wire is made of.

The students list the variables in their investigation:

- the length of the wire
- the thickness of the wire
- the material that the wire is made of
- the voltage of the battery

When you are choosing ideas to test, you need to make sure they can be tested by collecting data. All these questions involve variables that the students can change. They can control the other variables so that the data show the link between their chosen variable and the resistance of the wire.

A student might wonder if some bulbs look brighter than others to different people. This is an interesting question, but not one that you can collect data to test.

The students decide to investigate the length of the wire. They will measure the current that flows through different lengths of wire. This will show them how the resistance changes as they change the length. They need to keep the thickness of the wire, the material that the wire is made of, and the voltage of the battery the same.

Planning investigative work: preliminary work

The students need to work out which type of wire to use, which thickness of wire, and which voltage to use. They do some preliminary work to help them decide how to do the experiment safely, and how to get a good range of results.

They use secondary sources to find the most suitable wire. They try out different voltages for different lengths of wire.

Voltage (V)	Current for 10 cm of wire (A)	Current for 1 m of wire (A)
1.5		
3.0		
6.0		

They notice that the wire gets very hot for high voltages so they choose a voltage of 1.5 V.

Planning investigative work: choosing equipment

The students use a digital ammeter to measure the current in the wire. The ammeter will measure the current to the nearest 0.1 A.

They decide to change the length using clips to connect to different positions on the wire. They will use a metre ruler to measure the length of the wire between the clips. They can measure the length to the nearest 1.0 mm.

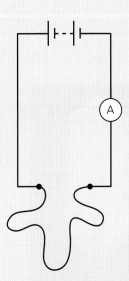

Planning investigative work: assessing hazards and controlling risk

The students know that the wire can get very hot. They put the wire on a heatproof mat and do not leave the wire connected to the battery for very long.

Obtaining and considering evidence

The students collect evidence by measuring the current for different lengths of wire. They repeat each experiment three times to make their results more reliable. Before calculating any averages they discard any anomalous results.

They plot a graph to try to find a pattern in their results. This will help them make a conclusion about the link between the current and the length of the wire.

Energy and power

Objectives

- Understand the difference between energy and power
- Be able to calculate power

The Kovalam lighthouse in Kerala, India has a very bright light at the top of a tall cylindrical tower. The light flashes 8 times every minute and can be seen nearly 40 km away. Ships use the light to navigate to the harbour at Kovalam.

Power

Some lamps are brighter than others. The lamp in the lighthouse is much, much brighter than the lamps used in torches or in houses or offices.

An electric current transfers energy from the **mains supply** to the lamp in the lighthouse. The lamp changes this energy into light and thermal energy.

A bright lamp gives out lots of light energy each second. It converts the energy from the mains supply into light energy faster.

The rate at which energy is converted is called the **power**. Power is measured in **watts** (W) or **kilowatts** (kW). 1 kW = 1000 W.

You can calculate power using an equation:

$$\text{Power (W)} = \frac{\text{energy transferred (J)}}{\text{time taken (s)}}$$

The lighthouse lamp changes 24 000 J of energy into light and thermal energy each minute.

The lamp in a torch changes the chemical energy in a battery into light and thermal energy. It transfers energy more slowly than the lighthouse lamp so the power is much less. The torch lamp transfers energy at a rate of 120 J each minute.

To calculate the power you have to convert the time to seconds.

For the lighthouse:

$$\text{Power} = \frac{24\ 000\ \text{J}}{60\ \text{s}}$$
$$= 400\ \text{W}$$

For the torch:

$$\text{Power} = \frac{120\ \text{J}}{60\ \text{s}}$$
$$= 20\ \text{W}$$

A lighthouse lamp is 20 times more powerful than a torch lamp.

Comparing light bulbs

For over 100 years most people have been using incandescent light bulbs in factories, homes, and torches. These lamps are the kind that you have used in electric circuits. The lamp has a **filament** which is very thin and glows when it gets hot.

In 1981 the first compact fluorescent lamp or CFL (energy-saving lamp) was sold. Since then, scientists have been developing a new kind of light. It is called a **light-emitting diode** or **LED** lamp. An LED is still powered by a battery or the mains but it is much more **efficient** than a normal light bulb. It converts far less of the energy into thermal energy and much more into light. The table

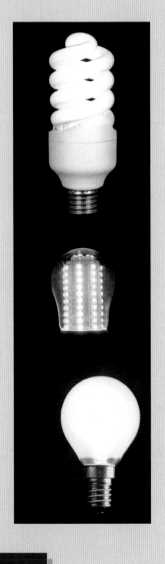

below shows how much power each type of lamp needs to produce different light levels.

Light intensity (lumens)	Power of incandescent light bulb (W)	Power of CFL (energy-saving lamp) (W)	Power of LED lamp (W)
450	40	10	7.5
800	60	15	10
1400	75	20	14
1800	100	25	18
2800	150	45	26

LEDs can be used in a huge variety of ways. They produce a lot of light for their size, and can be made small enough to sew into your clothes. Doctors use the intense light in cancer treatments, and LEDs were used in the Beijing and London Olympics to make spectacular patterns inside the Olympic stadium and on the outside of buildings.

Energy and money

When people pay their electricity bill they pay for energy, not current or voltage. The electricity supplier works out how much energy they have used. It depends on the power of their appliances and how long they are used for.

We can rearrange the equation for power like this:

Energy (J) = power (W) × time (s)

People use lots and lots of joules of energy each month. So instead the energy is measured in a different unit, the kilowatt-hour (kWh).

Energy (kWh) = power (kW) × time (h)

For example, if you use a 10 W (0.01 kW) LED for 10 hours in one month, the energy that you would pay for would be:

Energy = 0.01 kW × 10 h

   = 0.1 kWh

To produce the same amount of light using a normal light bulb you would need to use a 60 W light bulb. You would pay for 0.6 kWh instead. Using new light bulbs can save a lot of money. LEDs are a lot more expensive to purchase than incandescent or CFL light bulbs, but they do last a long time.

↑ Cancer treatment with LEDs.

Q

1 How many watts are there in 2 kW?

2 A CFL transfers 1200 J of energy each minute. What is the power of the lamp?

3 What are the advantages and disadvantages of LEDs?

4 Look at the calculations for the energy cost of using an LED and a normal bulb for 10 hours each month.

  a Calculate how much energy in kWh you would need for a CFL lamp to produce the same amount of light.

  b The electricity company charges 10 rupees per kWh. How much would it cost to use each lamp for 10 hours each month?

  c How much money would you save each month if you used an LED instead of an incandescent bulb?

!

- Power is the energy transferred per second.

- Power is measured in watts (W) or kilowatts (kW).

- You can work out the energy from power × time.

- LEDs have a wide range of uses.

1 Explain what is meant by the following words or phrases:

  a neutral [1]

  b charged [1]

  c charge [1]

  d earthed [1]

  e conductor [1]

  f insulator [1]

2 Copy and complete the sentences below using the words in question 1. You may need to use some words more than once.

  a In a thunderstorm air moves to produce regions that are _____. [1]

  b The air is usually an _____ but if there is a large amount of _____ in the cloud it can become a _____. [1]

  c When this happens _____ flows between clouds, or between clouds and the ground, and we see lightning. [1]

  d A lightning _____ is a piece of metal attached to a building. If it is struck by lightning the _____ flows through the metal. We say that the building is _____. [1]

3 Match each object to its correct definition. Write out the numbers and the letters. [4]

  Definitions:

  1 two plates with an insulator between them, that can be charged

  2 the insulator between two plates of a capacitor

  3 small squares that an image is made of

  4 the grid that absorbs light to make an image

  Words:

  A CCD

  B pixels

  C capacitor

  D dielectric

4 Look at the sentences below. Which circuit component or type of circuit does each sentence describe?

  a This is the energy source for an electric circuit. [1]

  b In this type of circuit all the components are in a single loop. [1]

  c This enables you to turn a component on or off. [1]

  d In this type of circuit all the components can be turned on or off independently. [1]

5 Look at the circuit diagram below.

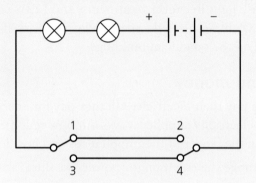

The circuit contains two special switches. The switch on the left can be flipped between connections 1 and 3. The switch on the right can be flipped between connections 2 and 4.

  a Copy and complete the table below to show what happens to the light bulbs when the switches are flipped. [1]

Position of switch on the left	Position of switch on the right	Are the bulbs on or off?
1	2	
1	4	
3	2	
3	4	

  b This circuit can be used to control the lights at the top and bottom of a staircase. Explain how it works. [1]

6 A student uses four identical bulbs and a battery to make a circuit. He puts a box over the circuit so the connections cannot be seen.

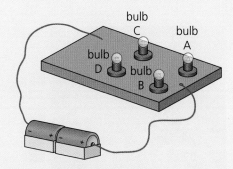

Another student unscrews each bulb in turn and watches the effect on the other three bulbs.

If you take this bulb out...	...this is what happens
A	bulbs B, C, and D go out
B	bulb C goes out; bulbs A and D stay on
C	bulb B goes out; bulbs A and D stay on
D	bulbs A, B, and C stay on

a Draw a diagram to show how the bulbs must be connected. [1]

b Which bulb or bulbs will be the brightest? Explain your answer. [2]

c Which bulb or bulbs will be the dimmest? Explain your answer. [2]

7 Here is a circuit diagram. All the lamps are the same.

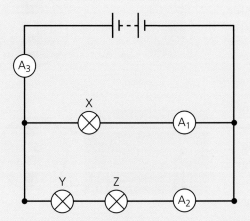

a Are the lamps connected in series, parallel, or both? Explain your answer. [1]

b Which ammeter will show the lowest reading? Explain your answer. [1]

c What is the link between the readings on the three ammeters? [1]

d Which lamp will be brightest? Why? [1]

e What would happen to the other lamps if: [3]

   i lamp X broke

   ii lamp Y broke

   iii lamp Z broke?

8 A circuit contains a 6 V cell and two lamps connected in series.

a Draw a circuit diagram showing how you would connect a voltmeter to measure the voltage across one of the lamps. [1]

b If both lamps are identical what will be the reading on each voltmeter? [1]

c Why? [1]

9 A student is planning how to collect some data to answer this question:

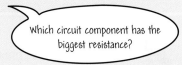

Which circuit component has the biggest resistance?

He writes down his ideas, but they are not in the right order. List the letters in order to put the statements in the right order.

A	Think about how I can make sure that the experiment is safe.
B	Connect up the circuit and measure the current through each component.
C	Select equipment so that I can measure the current.
D	Do some preliminary work to find the best voltage to use.
E	Write the results in a table.

[1]

10 A student wires up a circuit to measure the current through a buzzer and a lamp. He connects a voltmeter to measure the voltage across the buzzer. He draws his circuit diagram.

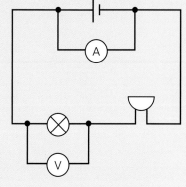

a How many things are wrong with his circuit? List them. [1]

b Draw the correct circuit diagram. [1]

203

Hot and cold

Thermal images

Objectives

- Explain the difference between temperature and thermal energy
- Describe how the motion of particles changes when you heat solids, liquids and gases

A scorpion is a cold-blooded animal, and the person holding it is warm-blooded. You cannot tell the temperature of the scorpion or the person by looking at them. The image of the scorpion is taken with special camera. It produces a **thermal image** that shows the differences in temperature.

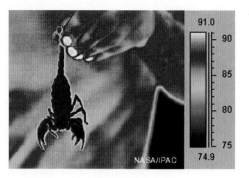

The difference between energy and temperature

The **temperature** tells us how hot or cold something is. We use a **thermometer** to measure temperature. A liquid inside a very narrow glass tube **expands** when it is heated.

For a long time thermometers were made using liquid mercury. Mercury is poisonous so today the liquid used in thermometers is alcohol. Digital thermometers do not use a liquid – instead they use a sensor to measure the temperature.

We measure temperature in **degrees Celsius** (°C).

- Human body temperature is 37 °C.
- The coldest temperature ever recorded on Earth was –89.2 °C, in Antarctica in 1983.
- The hottest temperature ever recorded was 56.7 °C in Death Valley, USA in 1913.

When we heat up a substance such as water, its temperature rises. We transfer energy from a chemical store of a fuel such as gas to the thermal store of energy in the water. Thermal energy and temperature are not the same.

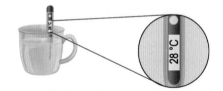

A cup of warm water and a swimming pool can have exactly the same temperature. However, the swimming pool stores much more thermal energy than the water in the cup. Unlike thermal energy, temperature does not depend on the amount of material.

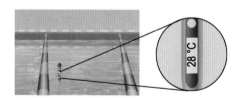

Heating solids, liquids, and gases

The process of heating changes the motion of particles. If you heat a solid the particles in the solid vibrate more. In liquids and gases the particles move faster. The temperature has increased.

You cannot say that the individual particles in a solid, liquid, or gas get hotter. Each particle can only move or vibrate faster. The temperature of a solid is a property of the solid, not of the individual atoms or molecules that make up the solid.

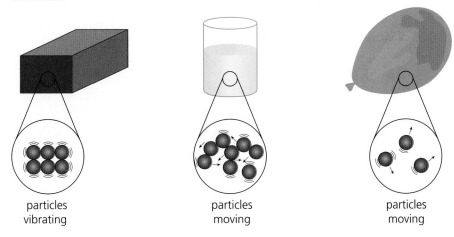

particles
vibrating

particles
moving

particles
moving

How much energy does it take to raise the temperature of an object? It depends on three things.

1. *The mass of the object*. In a greater mass there are more particles. You need to transfer more energy to get them all moving or vibrating faster.

2. *What it is made of*. The particles in different materials have different masses. If the particles are more massive you need more energy to get them moving or vibrating faster.

3. *The temperature change that you want*. For a bigger temperature change you need to get the particles all moving or vibrating even faster. This takes more energy.

Hot to cold

Thermal energy always moves from a hot object to a cold object. If you put a cold object in contact with a hot object the temperature of the cold object will increase, and the temperature of the hot object will decrease. Thermal energy has been transferred.

Look at the ice cubes in a glass of water. Thermal energy will be transferred from the water to the ice and it will melt. If you put a hot pan in a sink of water the water will get hotter. Thermal energy will be transferred from the pan to the water. Eventually the pan and the water will end up at the same temperature.

Q

1 Explain the difference between temperature and thermal energy.

2 Copy and complete these sentences.

When you heat up a liquid the particles in it move _____. The same thing happens when you heat up a _____. When you heat up a _____ the particles just vibrate more.

4 Why does it take more energy to heat up 1 kg of cold water than 0.5 kg of cold water to the same temperature?

5 Why does it take longer to boil a kettle of water than to warm the same kettle of water to a lower temperature?

- Temperature is a measure of how hot something is, measured in degrees Celsius.

- Thermometers are used to measure temperature.

- Temperature is independent of mass.

- When you heat up a solid, liquid, or gas the motion of the particles changes.

Energy transfer: conduction

Thermal energy

Neela wants to light a Bunsen burner. She takes a piece of wood, called a splint, and lights it from the teacher's Bunsen burner. She carefully carries the splint back to her Bunsen burner and lights it. The splint is on fire at one end, but she can hold it without burning her hand. This is because the thermal energy is not transferred down the splint to her hand.

Conductors and insulators

Thermal energy is transferred through a solid by **conduction**. Some solids conduct thermal energy better than others. Metals are very good conductors of thermal energy. Copper is a particularly good conductor of thermal energy, and saucepans are often made of copper. It quickly conducts the energy to the food in the saucepan.

Many non-metals are poor conductors of thermal energy. They are insulators. This does not mean that they do not conduct at all, but that thermal energy is transferred very slowly through them. Materials such as paper, cloth, wood, and plastic are all insulators. The handles of saucepans are often made of wood or plastic so you do not burn yourself.

Feeling warm or cold

Bicycle handlebars are made of metal but the ends of the handlebars are often covered in plastic. When you put your hand on the metal it feels cold, but the plastic does not feel cold.

Neela puts her hand on the metal of her handlebars, and then on the plastic.

Like Neela did, it is tempting to say that objects are at different temperatures if that is how they feel to us. However, our skin is not a very good thermometer. It detects the *transfer* of thermal energy rather than the temperature.

When Neela puts her hand on the metal, thermal energy is transferred away from her hand to the metal. She says it feels cold.

When Neela puts her hand on the plastic, thermal energy is not transferred away from her hand because the plastic is an insulator. She says it feels warm, though it is at the same temperature as the metal.

If you put your hand on something hotter than your skin, like a stone that is in the Sun, it feels warm. Thermal energy moves from the stone to your hand so it feels warm.

The metal is cold but the plastic is warm.

Conduction and particles

The diagram shows a piece of metal being heated at one end. The metal atoms on the left are vibrating a lot and the atoms on the right are not vibrating very much. The left end of the bar is hot and the right end of the bar is cold, so thermal energy is being transferred from the left to the right.

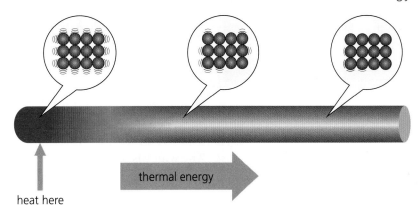

heat here

thermal energy

Conduction in liquids and gases

Liquids like water are poor conductors of thermal energy compared with solids. You can see this if you heat one end of a test tube of water. An ice cube held down by a weight will not melt, even if the water at the top is boiling.

A diver wears a wetsuit to dive in cold water. The wetsuit uses water as an insulator. It holds a small amount of water inside, forming a very thin layer between the diver's skin and the rubber of the suit. The water is not free to move around. Thermal energy is not easily transferred from the diver to the sea water so the diver does not feel cold.

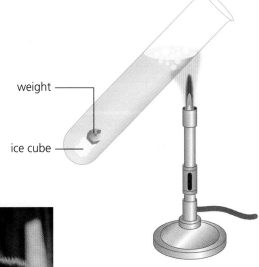

weight

ice cube

Air is a very good insulator if it is trapped and cannot move. It is used in materials that are designed to keep people warm. Clothing and bedding can be made from materials that trap air.

Gases such as air do not transfer thermal energy because their particles are much further apart than the particles in a solid.

Q

1 Describe the difference between a conductor and an insulator.

2 What would happen if you heated water in a saucepan made of a material that is not a good conductor of thermal energy?

3 Some divers use drysuits rather than wetsuits. They wear special clothing underneath that traps a layer of air between their skin and the suit.

   a Explain how a drysuit keeps a diver warm.

   b Do you think a wetsuit or drysuit keeps the diver warmer?

   c Explain your answer.

4 A student says that her blanket keeps her warm because it traps heat. What would you say to her?

!

- Thermal energy is transferred through a solid by conduction.

- Metals are good conductors of thermal energy, but plastic and wood are poor conductors.

- Liquids and gases that are not free to move around are poor conductors of thermal energy.

- Poor conductors are called insulators.

Energy transfer: convection

Objectives

- Describe what happens in convection
- Explain how convection currents are formed

Convection

When you heat the bottom of a saucepan of water, all of the water gets warm, not just the bottom part. The water is a liquid so it does not conduct thermal energy. The liquid in the pan is free to move so energy is transferred by **convection**.

The bottom of the pan is heated by the flame. Thermal energy is transferred through the pan to the water at the bottom of the pan by conduction. This is because the bottom of the pan is solid.

- The water in contact with the bottom of the pan gets warmer.
- The molecules in the warmer water are moving faster than the molecules in the cooler water above.
- The molecules in the warmer water move further apart.
- The warmer water becomes less dense.
- The warmer water rises (floats up).
- Cooler, denser water moves down to take its place.

Eventually all the water in the pan is circulating up and down. The water as a whole gets hotter and hotter and the water will **boil**. The circulation of water that is set up in this way is called a **convection current**. Convection currents can happen in all liquids and gases that are free to move.

Smoke from a fire rises because the air above the fire gets hotter, expands, and rises. If there is a fire in a house a firefighter may crawl along the floor to look for survivors. This is because there is less smoke near the floor than there is higher up.

Convection in the atmosphere

If you live near the coast you might notice that there is a breeze that comes off the sea during the day, but changes direction at night. We can explain this using convection.

On a hot day the Sun heats the land. The land is solid and heats up much more quickly than the water. The air just above the ground gets hotter than the air higher up. This hotter air expands and rises just like the hot water in a pan. The rising air is replaced with cooler air from above the sea, so the breeze comes off the sea.

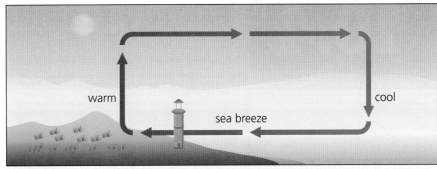

At night the ground cools down quickly, much more quickly than the sea. Now the air above the sea heats up, expands, and rises. This pulls cooler air from above the ground to replace it so the breeze changes direction.

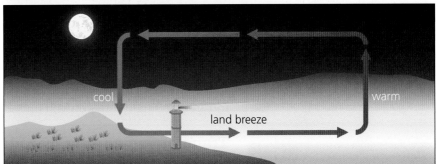

Convection also happens on a bigger scale. The Sun heats the Earth more at the equator than it does nearer the poles. The air near the equator rises and is replaced by cooler air. This causes the air in the atmosphere to move.

You might see some birds using the convection currents produced when the Sun heats the ground. The convection currents in the air are called **thermals**. In India, Africa, and Indonesia a small bird of prey called a shikra uses thermals to travel upwards to a great height. The bird can then swoop down to kill its prey.

Q

1 What is the difference between conduction and convection?

2 Explain what 'hot air rises' means, in terms of the particles in a gas.

3 This is a diagram of the Earth showing a convection current.

   a Is the direction of the convection current clockwise or anticlockwise?

   b Explain your answer.

4 Explain why there are no convection currents in solids.

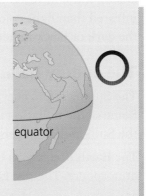

equator

!

- When a gas or liquid gets hot it expands and becomes less dense.

- The less dense gas or liquid rises and is replaced by cooler gas or liquid.

   This sets up a convection current.

- A convection current in the atmosphere is called a thermal.

Energy transfer: radiation

When you go outside you feel the warmth of the Sun. This is because the Sun emits radiation. Thermal radiation, **infrared radiation** or infrared, travels from the Sun to the Earth, just like light. Like light, infrared travels as a wave and can be reflected, transmitted, and absorbed.

All objects emit radiation, but the radiation that they emit depends on the temperature. The wire inside a torch bulb can get hot enough to give out infrared and light, but you will only ever emit infrared radiation.

Objectives

- Know some sources of infrared radiation and the similarities between light and infrared
- Describe how infrared is transmitted, absorbed, and reflected
- Explain what is meant by the greenhouse effect

Transferring energy by radiation

You have seen that thermal energy can be transferred by conduction and convection. Infrared radiation is another way that thermal energy can be transferred. Radiation is different from conduction and convection because it does not need a material for the thermal energy to be transferred.

Light and infrared reach the Earth from the Sun by travelling through space. There is no material in space. We say it is a **vacuum**.

Reflecting radiation

At the end of a marathon race the runners sometimes wrap themselves in sheets made of foil. The foil reflects the infrared back towards their bodies so that it doesn't escape. This helps to keep them warm because their body temperature drops quickly when they stop running.

Absorbing radiation

Thermal imaging cameras absorb infrared to produce an image, in the same way that a normal camera absorbs visible light to produce an image. Your skin detects infrared. When your skin absorbs infrared it gets warmer.

You saw on page 113 that objects appear black if they absorb most of the visible light that hits them. Objects also absorb infrared and some objects absorb more infrared than others. Dull, dark surfaces tend to absorb more infrared than light-coloured shiny surfaces. If you hang wet clothes out to dry, the dark or black materials will dry first.

The energy balance of the Earth

The Sun transfers energy to the Earth by radiation, such as light and infrared. Some of the radiation is reflected by the Earth's atmosphere, but about 50% of the Sun's radiation gets through and is absorbed by the land and oceans.

The Earth itself emits infrared and some of that is reflected back by the atmosphere. The rest is transmitted through the atmosphere into space.

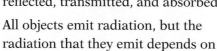

The temperature of the Earth depends on the balance between the radiation that is absorbed and the radiation that is emitted. If more is absorbed than is emitted then the temperature will increase.

The atmosphere reflects radiation because it contains **greenhouse gases**. These absorb infrared and then emit it again. This is the **greenhouse effect**. It is a good thing because without an atmosphere the Earth would be very, very hot in the daytime, and very, very cold at night.

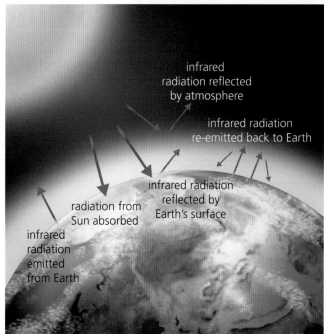

infrared radiation reflected by atmosphere

infrared radiation re-emitted back to Earth

infrared radiation reflected by Earth's surface

radiation from Sun absorbed

infrared radiation emitted from Earth

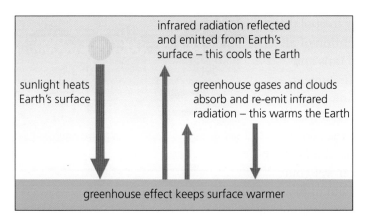

infrared radiation reflected and emitted from Earth's surface – this cools the Earth

sunlight heats Earth's surface

greenhouse gases and clouds absorb and re-emit infrared radiation – this warms the Earth

greenhouse effect keeps surface warmer

Greenhouse gases include carbon dioxide, methane, and water vapour. The concentration of greenhouse gases is increasing, and this is causing **climate change**. There are changes in the weather, icecaps are melting, and climate change affects how easy it is to grow food.

The electromagnetic spectrum

Light and infrared are just two of the types of radiation emitted by the Sun. They are two waves in the **electromagnetic spectrum**. Light and infrared are parts of the electromagnetic spectrum just like blue and green are two colours of the light spectrum.

radio	microwave	infrared	visible light	ultraviolet	X-rays	gamma
			ROYGBIV			

All the waves of the electromagnetic spectrum travel at the same speed, the speed of light, and they can all travel through a vacuum. They have different wavelengths, frequencies, and energies. Radio has the longest wavelength, the lowest frequency, and the smallest energy. Gamma has the shortest wavelength, the highest frequency, and the highest energy. They all have different uses and some are more dangerous than others.

Q

1 Look at the picture of the person holding the scorpion on page 204.

  a Why is the image of the person that we would see with our eyes different from the thermal image?

  b Why does the person not emit light?

2 If you stand near a fire on a cold night you feel warm even though the air is cold. Why?

3 Does the atmosphere trap heat? Explain your answer.

4 What is one consequence of climate change?

!

- Energy reaches us from the Sun as infrared or thermal radiation.

- Infrared can travel through a vacuum as a wave, like light.

- All objects emit thermal radiation. They emit light only if they are hot enough.

Cooling by evaporation

Cooling things down

In hot countries people use cooling by **evaporation** to cool down their houses. For thousands of years people have used water to cool their homes. Before the invention of air conditioning they used fans and water to cool a room. In ancient Egypt people would hang wet mats over the open doorway and use fans to cool down the air in the room.

You may have noticed this cooling effect if you wet your hands and leave them to dry.

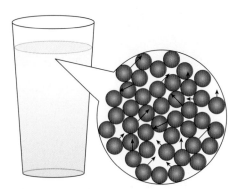

Inside a liquid

We can think about the molecules inside a liquid moving around in every direction. There is not very much space between the molecules and they are touching each other and moving over each other. Each molecule is attracted to the molecules around it. A molecule would need to travel fast to escape from the other molecules around it.

Why does water evaporate?

If you leave a dish of water out, it will **evaporate**. If you come back hours or days later the water will no longer be there. This will happen wherever you live as long as the temperature of the air is above 0 °C.

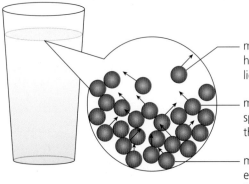

molecules at the surface with high speed escape from the liquid – evaporation

molecules with medium speed are pulled back to the liquid water

molecules with lower energy remain in the liquid water

The temperature of a liquid is linked to how quickly the molecules in the liquid are moving. The molecules do not all move at exactly the same speed. The molecules in the liquid have an **average speed** that depends on the temperature of the liquid. At a higher temperature the average speed will be higher. However, at any temperature some of the molecules will be moving slower than the average speed, and some will be moving faster. There is a range of speeds. This is because the molecules are colliding with each other and energy is transferred in those collisions.

In a liquid at a particular temperature, a few of the molecules will be travelling fast enough to be able to escape from the attraction of the other molecules around them.

When these fast-moving molecules escape, this reduces the average speed of the molecules left behind. You can think of it like this. Imagine a group of 10 children of different heights. To calculate the average height we add up all the heights and divide by 10. If the two tallest children leave the group, the average height will be smaller.

If the average speed of the molecules in a liquid is lower, this means that the liquid is cooler. The molecules are still moving around and there is still a range of speeds. A few will still be travelling fast enough to escape. So this will reduce the temperature of the liquid again.

Eventually all the molecules escape. The liquid has evaporated. The air around the liquid has cooled down.

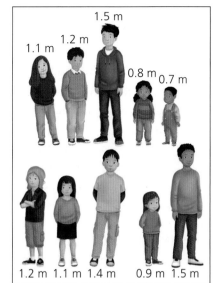

Average height =

$$\frac{(1.1 + 1.2 + 1.5 + 0.8 + 0.7 + 1.2 + 1.1 + 1.4 + 0.9 + 1.5)}{10}$$

= 1.14 m

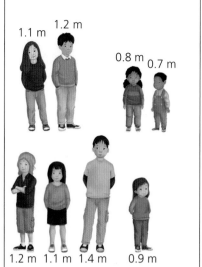

If the two tallest children leave the group then the average height =

$$\frac{(1.1 + 1.2 + 0.8 + 0.7 + 1.2 + 1.1 + 1.4 + 0.9)}{8}$$

= 1.05 m

Using cooling by evaporation

Cooling animals

Your body uses cooling by evaporation. When you get very hot you sweat. Droplets of sweat, which is mainly water, form on the surface of your skin. The water evaporates as thermal energy is transferred from your skin to the water. The thermal energy makes the water molecules move more quickly, so they can escape from the liquid water. You feel cooler because thermal energy is being transferred away from your body.

A dog will pant to help it to cool down. Water evaporates from its tongue. Elephants will use water on their skin to cool down.

Evaporative coolers and refrigerators

Some people use **evaporative coolers** instead of air conditioning to cool their house. In an evaporative cooler a fan blows warm air over water. Thermal energy is transferred to the water. It evaporates, and the air cools down. The cooled air circulates around the house.

In a **refrigerator** a special liquid called a **refrigerant** is pumped around tubes at the back. Thermal energy from inside the refrigerator evaporates the refrigerant. This cools down the inside of the refrigerator. Air conditioning works in the same way.

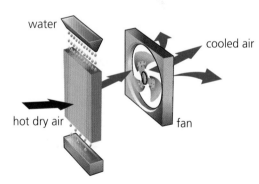

Q

1 A student thinks that water evaporates because it boils away. What would you say to him?

2 Why does water evaporate faster if the air around it is warmer?

3 Explain why your hands feel cool if they get wet.

4 Evaporative coolers need a supply of electricity to work. Why?

!

- Liquids evaporate when fast-moving molecules escape.
- Evaporation cools the liquid.
- Evaporative coolers transfer energy from air to a liquid to cool the room.

The world's energy needs

Primary and secondary energy sources

You often hear people talking about the world's energy needs. They are talking about the **primary energy sources** (or resources) like coal and **secondary energy sources** like electricity.

Primary energy sources	Primary energy converters	Secondary sources/ carriers
coal	power stations (e.g. coal)	electricity
oil	generators (e.g. wind)	fuels such as petrol/gasoline
natural gas	solar cells	hydrogen gas
sunlight	refineries (to produce petrol/gasoline)	
uranium	hydrogen gas production	
wind		
water (tides, lifting water up)		
biofuel (wood, waste)		
geothermal (using thermal energy from under the ground)		

Some primary sources can be found in the ground, such as coal, **oil**, **gas**, or **uranium**. Other sources like **wind**, **water**, and **biomass** are in the environment around us.

Some primary sources such as coal can be used directly to heat homes or cook food. Coal is also used to generate electricity in a **power station**. Electricity is an example of a **secondary source** because it has been produced using a primary source. **Energy converters** such as **power stations** produce secondary sources such as electricity. **Refineries** are energy converters that produce other fuels from oil, such as petrol (gas or gasoline).

Renewable and non-renewable energy sources

Some energy sources such as coal, oil, gas, and uranium will run out. We say that they are **non-renewable**. We cannot get more of them during our lifetime. Other sources, such as wood, are **renewable**. We can grow more trees if we need more wood. Some people think that renewable energy resources can be used again, but that is not correct. 'Renewable' means you can replace it fairly quickly.

How are the world's energy needs changing?

150 years ago most people in the world were using primary energy sources that they could cut down or dig up, like wood or coal.

The chart shows how the world's energy needs have been met in the last 150 years. It shows the energy in **exojoules** (EJ). One exojoule is one million, million, million joules.

There has been a huge increase in the use of non-renewable energy sources like coal, oil, gas, and uranium (nuclear), but the use of biomass such as wood has stayed roughly the same. The use of renewable sources like wind and water is very small (the green line on top of the brown band).

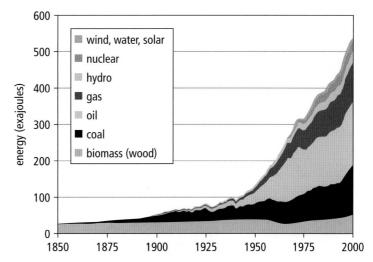

⬆ Primary energy sources used throughout the world.

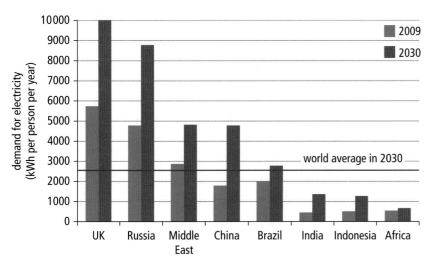

⬆ Predicted demand for electricity by some countries of the world.

This chart shows predictions for the demand for electricity in 2030. A kWh is another unit of energy.

1 kWh = 3 600 000 J

Both charts show **secondary data**. The data were not collected by you in a laboratory, but collected and displayed by other people. Data you collect yourself are **primary data**.

Q

1 There are primary and secondary energy sources.

  a Explain the difference.

  b Is petrol (gas or gasoline) a primary or a secondary source? Why?

  c Is oil a primary or a secondary source? Why?

2 Look at the global energy supply.

  a Which energy source was used the most in 1900?

  b Which energy source was used the most in 2000?

  c By how much has energy supply increased between 1900 and 2000? Choose from these options: it doubled, it trebled, it is ten times bigger.

3 Explain the difference between a renewable and a non-renewable source of energy.

4 Look at the electricity demand chart. Which country or countries will double or more than double their demand for electricity between 2009 and 2030?

- Primary energy sources are coal, oil, gas, and uranium.

- Secondary energy sources are electricity and fuels such as petrol/gasoline and hydrogen.

- Non-renewable resources will run out, but renewables will not.

- The demand for primary energy sources needed to generate electricity is increasing rapidly.

Fossil fuels

Fossil-fuel power stations

Most power stations use the stored energy in fossil fuels to generate electricity. Fossil fuels are non-renewable energy sources so they will run out. We need to work out how to meet our electricity needs in the future.

The diagram below shows how a **coal-fired power station** works. Oil- and gas-fired power stations work in the same way. Power stations need to be built near a water supply such as the sea, a river, or a reservoir.

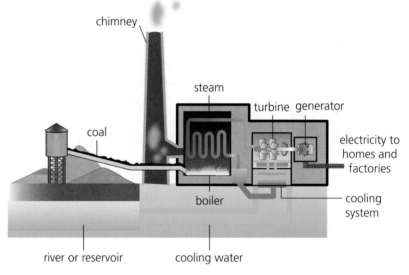

- In the power station the coal, oil, or gas is burned to turn water into steam.
- The steam drives a **turbine**, which is like a giant fan that spins round.
- The turbine is connected to the **generator**. This produces a voltage in the coils around it.
- The steam is turned back to water in the condenser.

You will learn more about how electricity is generated on pages 218–9.

Burning fossil fuels in power stations produces carbon dioxide, which is a greenhouse gas. Large amounts of carbon dioxide can affect the climate. Burning fossil fuels can also pollute the air. You can find out more about pollution in your *Complete Chemistry for Cambridge Secondary 1* Student book. However, fossil-fuel power stations are a very reliable source of electricity that produce electricity all the time.

Most of the energy wasted in generating electricity is thermal energy. You could say that a power station is better at generating thermal energy than generating electricity. In some power stations that thermal energy is used to heat water, which heats nearby homes and factories.

How were fossil fuels formed?

Fossil fuels formed from plants and animals that lived millions of years ago.

Millions of years ago trees grew near swamps. When they died they were covered by mud and sand. The trees did not rot because there was no oxygen. Mud and sand piled up and compressed the dead trees. Thermal energy from beneath the Earth and the pressure of the mud changed the trees into coal.

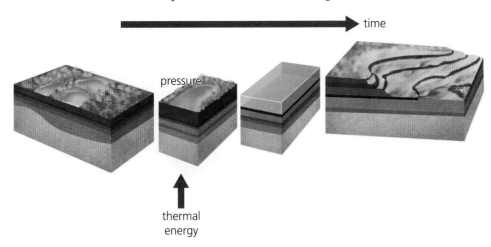

Millions of years ago the sea was full of small sea creatures with shells made of calcium carbonate. When they died they were covered by mud and sand. Mud and sand piled up and compressed them. The bodies of the sea creatures were turned into oil and gas by the pressure and thermal energy.

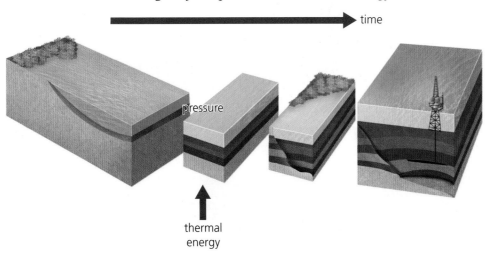

Scientists are developing new ways to find fossil fuels underground. They can only make an estimate of the amount of coal, oil, and gas there is in the world.

Q

1 Copy and complete these sentences.

When you burn coal in a power station the thermal energy is used to turn _____ into _____. This drives a _____, which drives a generator to produce electricity.

2 Why do you need to build a fossil-fuel power station near a river or the sea?

3 Describe one similarity in the formation of coal and of oil. Describe one difference.

4 Why are coal, oil, and gas called *fossil* fuels?

!

- In a fossil-fuel power station the fuel is burnt to turn water into steam.

- Steam turns the turbine, which turns a generator to generate electricity.

- Burning fossil fuels produces carbon dioxide.

- Coal, oil, and gas are the remains of plants and animals that lived millions of years ago.

217

10.8 Generating electricity

What do we mean when we say that 'electricity is generated'? We know that to make a component work in a circuit we need a voltage. Generating electricity means producing a voltage so that a current can flow.

Inside a generator

You saw in Chapter 7: Magnetism on page 140 when a current flows in a wire it produces a magnetic field around the wire. There is a link between electricity and magnetism.

Bulan takes a coil of wire and connects it to a voltmeter.

I wonder how electricity is produced in a power station.

If she moves the magnet towards the coil there is a positive reading on the voltmeter.

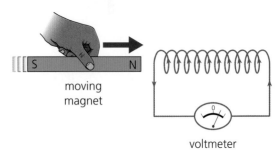

moving magnet

voltmeter

If she holds the magnet still the reading on the voltmeter is zero.

magnet not moving

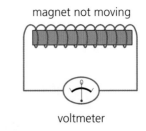

voltmeter

If she moves the magnet away from the coil there is a negative reading on the voltmeter.

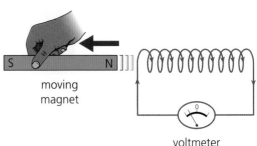

moving magnet

voltmeter

We say that a voltage has been produced or **induced** in the coil of wire. If you replace the voltmeter by a lamp or other circuit component, a current will flow in the circuit.

Bulan tries the experiment again but this time she holds the magnet still and moves the coil of wire towards and away from the magnet. She sees the same thing happening on the voltmeter. If she moves the magnet towards the coil the reading is positive, but if she moves it away the reading is negative.

She realises that to induce a voltage you need movement, but it does not matter if the magnet is moving or the coil is moving.

Inducing a bigger voltage

Bulan's friend Utari says that she has worked out how to make the reading on the voltmeter bigger.

Utari is correct. To make the induced voltage of her simple generator bigger.

She can:

- move the magnet (or the coil) faster
- use a coil with an iron core
- use more turns on the coil
- use a stronger magnet.

I think that the reading will be bigger if you move the magnet faster.

A simple generator

The simplest example of a generator is a bicycle **dynamo**. The dynamo uses the motion of a bicycle tyre to light the lamps on a bicycle.

turning bicycle wheel makes this wheel turn

magnet turns inside coils of wire

the induced voltage across the wire lights the lamps

Power station generators

The generators inside a power station are very, very big. They use moving electromagnets instead of permanent magnets. This is because electromagnets can produce a much stronger magnetic field than a permanent magnet, so the induced voltage will be bigger.

Q

1 What do we mean by an 'induced voltage'?

2 What are the main differences between a bicycle dynamo and a power station generator?

3 Utari moves a magnet into her coil of wire with the north pole first. This induces a voltage and the voltmeter reading is positive. She turns the magnet around so that the south pole is first. What will happen when she moves the magnet into the coil now?

4 Many people use battery-powered lights on their bicycles instead of dynamos.

   a What is one advantage of using a bicycle dynamo instead of a battery-powered light?

   b What is one disadvantage of using a bicycle dynamo instead of a battery-powered light?

!

- When a magnet moves near a coil of wire, or a coil of wire moves near a magnet, a voltage is induced.

- Large-scale generators are made using very large coils and powerful electromagnets.

Renewable energy: solar and geothermal

Objectives

- Describe how the energy from the Sun can be used
- Explain how energy from the Earth can be used to generate electricity

This is the Mithapur solar power station. There are over 100 000 solar cells covering 0.4 square kilometres. India has about 300 sunny days a year and could generate all the electricity it needs using solar cells.

Solar cells

A **solar cell**, or photovoltaic cell, converts the energy of light directly into electricity. Unlike a generator, a solar cell has no moving parts. Each solar cell on its own does not produce a big voltage so you need to connect lots of solar cells together.

Solar cells have become more and more efficient since they were first made. In the 1970s solar cells had an efficiency of about 5% but this has increased to about 30%. This means that you do not need as many solar cells to produce the same amount of electricity.

As well as using solar cells in large power stations people can use them to produce electricity for their homes. Wherever you live, however remote, solar cells can generate electricity. Solar cells work even if it is cloudy, but they do not work in the dark.

With your own solar cells you can generate enough electricity for lighting and for electronic items like phone chargers and radios, but not usually enough if to heat your house. Solar cells are expensive to produce because of the cost of the materials. Although they do not produce greenhouse gases when you use them, greenhouse gases were produced when they were made.

Solar water heating

Solar water heating is different from solar cells.

A solar water heater consists of pipes inside a panel. The pipes are connected to a tank so that cold water flows into the pipes. It gets heated by the Sun and hot water flows out. The pipes are usually made of copper and are painted black.

A solar water heating system can be used to heat a hot water tank which is also heated by an electrical heater if necessary. Sometimes the heater is called a solar panel, but that name can be confused with the solar cell.

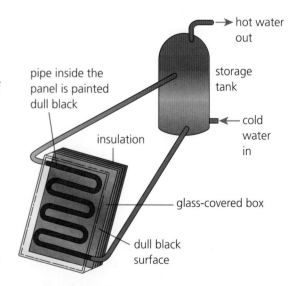

pipe inside the panel is painted dull black

insulation

hot water out

storage tank

cold water in

glass-covered box

dull black surface

Geothermal power

The centre of the Earth is very hot. You can find out more about the structure of the Earth in your *Complete Chemistry for Cambridge Secondary 1* Student book. Some of the energy was there when the Earth was formed and some of it comes from nuclear reactions in the Earth. If you drill down into the Earth the temperature rises about 25 °C for every kilometre closer to the centre of the Earth. This temperature rise can be used to generate electricity.

If you pump water down into the ground far enough it is heated enough to be converted into steam. That steam can be used to drive a turbine which drives a generator. Geothermal power stations do not produce much carbon dioxide while they are running, but they are very expensive to build. Building them produces lots of carbon dioxide.

The Olkaria II geothermal power station is one of Kenya's largest geothermal power stations. Pipes are drilled about 4.5 km down into rock.

Some people use the thermal energy from the Earth to heat their houses in the winter, or cool them down in the summer, by using a **heat pump**. This pumps water through pipes in the ground. The water gains thermal energy which can be transferred to heat the house. In the summer the ground is cooler than the house so the same system can be used to cool the house down.

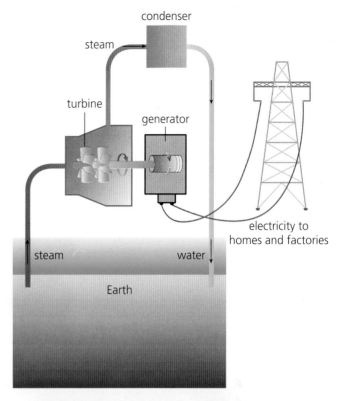

⬆ Pipes are buried deep in the rock below this geothermal power station.

Q

1 Look at the photo of the solar power station. Why do you need to connect lots of solar cells together?

2 What percentage of energy is wasted in a modern solar cell?

3 How many times more efficient is a modern solar cell compared with an old one?

4 What is the source of geothermal energy?

5 Look at the coal-fired power station on page 216. Compare it with the geothermal power station above.

   a What are the similarities between the two power stations?

   b What are the differences?

6 The power output of the Mithapur solar power station is about 25 MW. The power output of the Olkaria II power station is about 100 MW.

   How many solar power stations (roughly) would produce the same power as the Olkaria II geothermal power station?

!

● Solar energy can be used to produce electricity directly using solar cells.

● The energy from the Sun can also be used to heat water.

● Energy from deep inside the Earth can be used to turn water into steam. The steam can drive a turbine, which can drive a generator to generate electricity.

Renewable energy: using water and wind

Hydroelectricity

Hydroelectricity is another method of generating electricity that does not use fossil fuels. The photo shows Indonesia's largest hydroelectric power station, Cirata, 100 km south-east of Jakarta. It was completed in 1998 and its dam is 125 m tall.

In hydroelectricity it is liquid water rather than steam that moves a turbine. The turbine turns a generator to generate electricity.

A river fills up a big lake or **reservoir** behind a dam. To generate electricity the sluice gates are opened to allow water from the reservoir to flow down pipes inside the dam. The water flowing in pipes drives the turbines.

The water in the reservoir is a store of gravitational potential energy (GPE). The GPE changes to the kinetic energy of the water and turbines. The generator generates electricity which is used to power homes and factories.

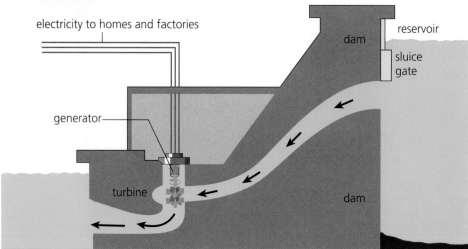

You need a suitable place to build a hydroelectric power station. Regions with mountains and lakes provide ideal locations. Sometimes valleys are flooded to produce a lake and this can affect the environment.

This type of power station is very expensive to build, and you need to connect the power station to houses and factories. However, you can produce electricity whenever you need to. You just need to open the sluice gates.

Wind

A **wind turbine** is a large windmill that is used to generate electricity. Although it is called a wind turbine it contains a turbine *and* a generator. Groups of wind turbines make **wind farms**. A wind turbine can produce a power of about 2 MW.

The Tamil Nadu wind farm in India produces just under 6000 MW from its collections of wind turbines.

The output of a wind turbine is not always predictable because the wind does not always blow. Some people like the way that wind turbines look, but others do not. They can be noisy and can kill birds. They are also expensive to build.

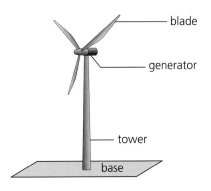

Wave power

Wave power works like wind power. There is a chamber on the shoreline. The waves force water up and down inside it. The motion of the water forces air backwards and forwards through a turbine, which makes a generator spin to generate electricity.

Wave power does not produce a constant supply of electricity. It depends how often the waves come. The chamber needs to be very strong because waves can be very powerful. You need to find a suitable site to build the chamber.

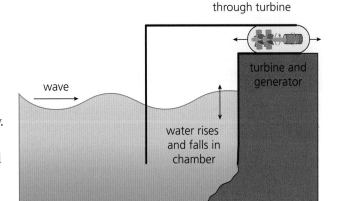

Tidal power

The largest **tidal power** station in the world is the Sihwa Lake tidal power station in South Korea. In 1994 a huge sea wall was built that contained ten large underwater turbines. This is sometimes called a 'barrage'. As the tide comes in the turbines spin and they drive generators to generate electricity. They spin again when the tide goes out. Tidal power is a bit like hydroelectric power.

There are tides every day so a tidal power station can produce electricity for about 10 hours each day. Some people are concerned that they can affect the environment. For example fish cannot swim upriver to breed. Tidal barrages are very expensive to build. Like hydroelectricity and wave power, greenhouse gases are produced while they are being constructed but not while they are running.

Q

1 What are the similarities between tidal power and hydroelectric power? What are the differences?

2 Are wind power and wave power 'free'? Explain your answer.

3 All methods of generating electricity have advantages and disadvantages.

   a What are the main advantages and disadvantages of wind power?

   b What are the main advantages and disadvantages of hydroelectric power?

4 Do any of these methods of generating electricity produce greenhouse gases such as carbon dioxide? Explain your answer.

- You can use water in lakes, wind, waves, and tides to generate electricity.
- Moving air or water is used to drive a turbine, which drives a generator.
- All of these methods are renewable.
- There are advantages and disadvantages to all of these methods.

Energy for the future

Demand for electricity

Since the first electric light bulb was turned on, and the first car engine was made, the demand for electricity and fuels has been increasing. It will continue to increase in the future. Up until now we have relied on fossil fuels to provide the petrol (gas or gasoline) for our cars, and the energy from burning fossil fuels to turn water into steam in our power stations. However, fossil fuels will run out. It is difficult to predict how long they will last. We do not know exactly how much coal, oil, or gas there is underground.

We need to plan how we will generate electricity in the future, and what we will use for transport.

- Could we use **electric cars**? Electric cars use large batteries rather than burning fossil fuels.
- Should we use **hybrid cars**? Hybrid cars can run on either electricity or fossil fuels.

Which is the best source of energy for the future?

Here are some questions to consider when looking at alternatives to fossil fuels.

Payback time

India uses a lot of renewable energy sources to generate electricity. Renewable energy sources such as wind and solar can be used on a large scale or by families in their homes.

Rohit Khandelwal lives a town near Jaipur. He wants to use renewable energy sources to reduce the cost of his electricity bill. He looks at the cost of a small wind turbine. The turbine would cost about 100 000 rupees. It would save him 10 000 rupees per year on his electricity bill. He uses this information to work out the **payback time**.

$$\text{Payback time} = \frac{\text{cost of turbine}}{\text{saving per year}}$$
$$= \frac{100\ 000\ \text{rupees}}{10\ 000\ \text{rupees/year}}$$
$$= 10\ \text{years}$$

Rohit will have paid back the cost of the turbine after 10 years. After that the electricity will be 'free'. However, the turbine will not last forever so he would have an advantage for about 10 years. After that he would probably need to buy another turbine.

Renewable sources

We are using more and more renewable fuels. Each country in the world is considering how to make the best use of renewable sources of energy. A government needs to consider the cost to the environment and other costs that a family would not worry about. For example, there is an energy cost to building turbines. There is also a cost in terms of pollution and greenhouse gases, which contribute to climate change, when you build turbines.

Biofuels

Biofuels are renewable energy sources that we get from plants or animals. These fuels have been used for thousands of years. Wood is one of the most common fuels throughout the world. However, there are lots of different biofuels that we can burn. **Biodiesel** is a fuel made from plant oils, and **bioethanol** is made from sugar. Plant material used for fuels is called **biomass**. Wood fuel, rubbish, alcohol fuels, crops, and landfill gas are all biomass.

All sources of energy have a cost. The energy from biofuels is not free. You need to buy a furnace to burn the fuel and a generator to generate electricity.

You could also buy a **biogas** digester. This uses the gas that is produced from rotting waste to generate electricity.

Fuels of the future

New technology uses **hydrogen** fuel to provide a source of energy for cars. **Hydrogen cars** do not produce the polluting gases such as nitrogen oxides that conventional cars do. The only gas that they produce is water vapour. Some cars can run on either hydrogen fuel or petrol (gas or gasoline). Others use **fuel cells** that make electricity by combining hydrogen and oxygen. One of the problems with hydrogen as a fuel is that there are not very many hydrogen fuelling stations in the world.

Nobody knows exactly how we will meet the demands for energy, but the way we generate electricity and move around could be very different to the ways that we do so today.

1 Is biomass a renewable fuel? Explain your answer.

2 Why do you not see many cars powered by hydrogen on the roads?

3 A family are looking at using a biomass generator to generate electricity.

　a Explain what is meant by payback time.

　b The cost of the generator is 24 000 rupees and the family would save 3000 rupees per year. Work out the payback time.

4 Why is it very difficult to decide which is the best source of energy to use?

- All renewable sources of energy cost money.

- You can work out the payback time to help decide the best option.

- There is also a cost in terms of energy, greenhouse gases, and pollution.

Review
10.12

1 Copy and complete these sentences.

  a You measure _____ in degrees Celsius. [1]

  b You measure _____ in joules. [1]

  c A bath will have _____ thermal energy than a cup of warm water at the same temperature. [1]

2 The amount of energy that you need to raise the temperature of something does not depend only on the temperature change.

  a What other two things does it depend on? [2]

  b List the letters in order from the smallest to the largest amount of energy required. The energy needed to raise the temperature of:

    A 1 kg of water by 20 °C

    B 2 kg of water by 20 °C

    C 1 kg of water by 10 °C [3]

3 Explain why:

  a saucepans are usually made of metal [1]

  b black clothes dry quicker than white clothes if they are drying in the sun [1]

  c a breeze comes onto the land during the day [1]

  d a bird plumps up its feathers if it is cold. [1]

4 A student puts two thermometers on the desk underneath a lamp. She covers the bulb of one thermometer with black paper and the other with aluminium foil.

  a Which thermometer will read the higher temperature after half an hour? [1]

  b Why? [1]

5 There is a hot drink on the table.

  a Explain how conduction changes the temperature of the tea. [1]

  b Explain how convection changes the temperature of the tea. [1]

  c Explain how radiation changes the temperature of the tea. [1]

  d Explain why putting a lid on top of the cup would keep the tea hotter for longer. [1]

6 a Explain what is meant by the greenhouse effect. [1]

  b Why is the greenhouse effect a good thing? [1]

  c Why is the greenhouse effect a problem? [1]

7 A student has left his drink outside in the sun.

  a Explain why the can of soft drink is warmer than it would be if it was inside the house. [1]

  The student decides to use the sun to cool down a can of drink on a hot day. He covers the can of drink with a pot and pours water over the pot. The water sinks into the pot and makes it wet. He keeps pouring water over it for a while. When he takes the can out it is cooler than before.

  b Explain why the can is cooler. [1]

  c Explain why he has to keep pouring water over the pot. [1]

8 Here is a list of energy sources.

  A coal   E uranium   I biomass

  B sunlight   F wind   J waves

  C oil   G gas   K electricity

  D hydroelectricity   H tidal   L geothermal

  a Which letters are renewable sources of energy? [1]

  b Which are non-renewable sources? [1]

  c Which can you burn to produce steam? [1]

d Which do you dig up from underground? [1]

e Which ones involve water? [1]

f Which one is the odd one out, and why? [1]

9 Decide whether each statement about generating electricity is true or false. [5]

a If you hold a magnet near a coil of wire a voltage will be induced.

b You move a magnet out of a coil and the induced voltage is positive, so when you move it in it will also be positive.

c It doesn't matter whether you move the coil or move the magnet to induce a voltage.

d Moving the magnet faster will induce a bigger voltage.

e The strength of the magnet doesn't matter, only the speed that you move it.

10 Here is a diagram of a fossil-fuel power station.

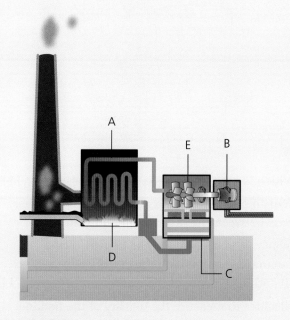

Which letter is the:

a turbine [1]

b generator [1]

c steam [1]

d fuel [1]

e water? [1]

f Which things would you also find in a geothermal power station? [1]

g Which things are needed to generate electricity using the wind? [1]

11 Look at the list of energy sources in question **8**. Here is a list of definitions, advantages, or disadvantages. For each one write down which energy source or sources matches each statement. You may need to use the same source twice.

a It uses water stored behind dams. [1]

b It contributes to climate change by producing carbon dioxide. [1]

c Building it can cause damage to the environment by flooding. [1]

d It can be unreliable because it depends on the weather. [1]

e It produces electricity on demand or all the time. [1]

12 Here are some secondary data about solar cells from a company that sells solar cells. The table shows the area of the panels, the power produced, the energy produced per year, and the mass of carbon dioxide that would be saved because you are not burning fossil fuels.

Area (m²)	Energy produced per year (kWh)	Mass of carbon dioxide saved per year (kg)
1	150	12
2	300	24
3	450	36
4	600	48

a A family want to save 6000 kWh per year. What area of solar panel would they need? [1]

b How much carbon dioxide would you save per year if you used 25 m² of solar panels on your roof? [1]

c Do you think that the energy produced as shown in the table is the maximum or the minimum energy that you would get? Explain your answer. [2]

1 A car is at the garage waiting to be repaired. The weight of the car is 20 000 N and the area of each tyre in contact with the ground is 250 cm².

a Calculate the pressure that the car exerts on the ground. [2]

b The tyres are a bit flat. The mechanic pumps them up so that there is less area in contact with the ground. Is the pressure now bigger or smaller? Explain your answer. [2]

The mechanic needs to lift the car up. Here is a diagram of a hydraulic pump. He can put a car on the ram and use the handle to lift the car.

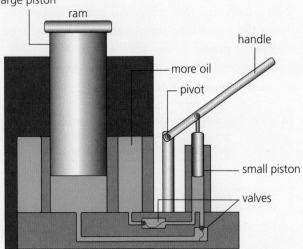

large piston
ram
handle
more oil
pivot
small piston
valves

The mechanic pushes the handle down and the oil pushes the large piston up. The valves let more oil in so that he can pump the handle again.

c What would happen if the pump used air instead of oil? [1]

d Why does it need to let more oil in each time? [1]

e The area of the large cylinder is 100 cm². What pressure will he need in the oil to lift the car? [2]

f The area of the small piston is 5 cm². What force does he need to apply to the small piston to produce that pressure? [2]

To make pumping the car up easier, the handle is part of a lever.

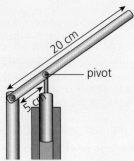

20 cm
5 cm
pivot

g What is a lever? [1]

h in part **f** you worked out the force that the mechanic needs to apply to the piston. What force does the mechanic need to apply to the handle? [2]

2 A girl is playing with some oil and water. She pours the oil and water into a glass and the oil floats on the water.

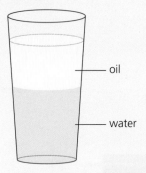

oil
water

a Which is more dense, the oil or the water? [1]

She cuts a cube of banana and puts it in the glass. It floats between the layer of oil and water.

b What can you say about the density of the banana compared with the density of the oil and the density of the water? [1]

She decides to measure the density of the banana.

c What equipment will she need to measure it? [1]

d Explain how she could use that equipment to measure the density. [3]

3 A teacher is showing the class how you can make a balloon stick to the wall by rubbing it on your clothes.

a As she rubs the balloon on her clothes the balloon becomes positively charged. What has happened to make the balloon charged? [1]

b The wall is neutral. Explain what this means. [1]

c As she brings the positively charged balloon towards the wall the surface of the wall becomes negatively charged. Explain why. [1]

d She leaves the balloon stuck to the wall but it eventually falls off. Why? [1]

e A student goes home and tries to stick the balloon to the metal case of a refrigerator. Will it work? Explain your answer. [2]

4 A student has connected a parallel circuit.

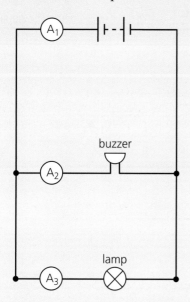

a The reading on ammeter A_1 is 0.5 A, and the reading on ammeter A_2 is 0.1 A. What is the reading on ammeter A_3? [1]

b If you swapped the components over, what would happen to the readings on the ammeters? [1]

c What would happen if you used a battery with a bigger voltage? Why? [1]

The student replaces the buzzer with a motor. The reading on ammeter A_2 is now 0.2 A.

d What is the reading on the other two ammeters? [1]

e Where would you put a voltmeter to measure the voltage across the motor? [1]

f Where would you put a voltmeter to measure the voltage across the battery? [1]

g Would the two voltmeter readings be the same or different? Explain your answer. [2]

h What would happen to the reading on ammeter A_3 if you added another lamp in that branch? Explain your answer. [2]

5 During the day in the desert it can get very hot. Some people run marathons in the desert. There is an 'ultra-marathon' that takes place in the Sahara desert.

a The temperature in the Sahara can reach 45 °C. What is the difference between temperature and thermal energy? [1]

b How does energy from the Sun reach a runner in the desert? [1]

c A runner will perspire (sweat) when he gets hot. Explain how that helps to keep him cool. [2]

d At the end of the race he goes inside a cool room. It feels cold. In which direction is thermal energy being transferred: from the room to him, or from him to the room? [1]

e In the cool room, is thermal energy transferred by conduction, convection, or radiation, or more than one of these methods? Explain your answer. [2]

6 A farmer wants to use solar power to reduce the cost of his electricity. He buys a solar cell and puts it on his roof.

a Is solar power renewable or non-renewable? Explain your answer. [2]

b Give one advantage and one disadvantage of using the solar cell rather than electricity generated in a coal-fired power station. [2]

c How would you expect the output of the solar cell to change during the day? Why? [2]

His neighbour decides to use a biogas generator instead. The biogas from waste is burned and this produces electricity.

d Explain the process that produces electricity from burning a fuel. [2]

e Why are some fuels called fossil fuels? [1]

f The neighbour says that we should all use only renewable sources of energy to generate electricity. Do you agree? Explain your answer. [2]

Choosing apparatus

There are many different types of scientific apparatus. The table below shows what they look like, how to draw them, and what you can use them for.

Apparatus name	What it looks like	Diagram	What you can use it for
test tube			• heating solids and liquids • mixing substances • small-scale chemical reactions
boiling tube			• a boiling tube is a big test tube. You can use it for doing the same things as a test tube.
beaker			• heating liquids and solutions • mixing substances
conical flask			• heating liquids and solutions • mixing substances
filter funnel			• to separate solids from liquids, using filter paper
evaporating dish			• to evaporate a liquid from a solution
condenser			• to cool a substance in the gas state, so that it condenses to the liquid state

stand, clamp, and boss			• to hold apparatus safely in place
Bunsen burner			• to heat the contents of beakers or test tubes • to heat solids
tripod			• to support apparatus above a Bunsen burner
gauze			• to spread out thermal energy from a Bunsen burner • to support apparatus, such as beakers, over a Bunsen burner
pipette			• to transfer liquids or solutions from one container to another
syringe			• to transfer liquids and solutions • to measure volumes of liquids or solutions
spatula			• to transfer solids from one container to another
tongs and test tube holders			• to hold hot apparatus, or to hold a test tube in a hot flame

Working accurately and safely

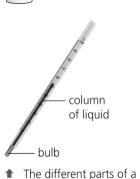

Using measuring apparatus accurately

You need to make accurate measurements in science practicals. You will need to choose the correct measuring instrument, and use it properly.

Measuring cylinder

Measuring cylinders measure volumes of liquids or solutions. A measuring cylinder is better for this job than a beaker because it measures smaller differences in volume.

To measure volume:

1. Place the measuring cylinder on a flat surface.

2. Bend down so that your eyes are level with the surface of liquid.

3. Use the scale to read the volume. You need to look at the bottom of the curved surface of the liquid. The curved surface is called the **meniscus**.

Measuring cylinders measure volume in cubic centimetres, cm³, or millilitres, ml. One cm³ is the same as one ml.

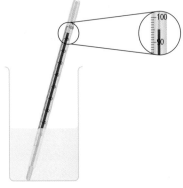

— column of liquid

— bulb

↑ The different parts of a thermometer.

Thermometer

The diagram to the left shows an alcohol thermometer. The liquid expands when the bulb is in a hot liquid and moves up the column. The liquid contracts when the bulb is in a cold liquid.

To measure temperature:

1. Look at the scale on the thermometer. Work out the temperature difference represented by each small division.

2. Place the bulb of the thermometer in the liquid.

3. Bend down so that your eyes are level with the liquid in the thermometer.

4. Use the scale to read the temperature.

Most thermometers measure temperature in degrees Celsius, °C.

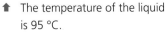

↑ The temperature of the liquid is 95 °C.

Balance

A **balance** is used to measure mass. Sometimes you need to find the mass of something that you can only measure in a container, like liquid in a beaker. To use a balance to find the mass of liquid in a beaker:

1. Place the empty beaker on the pan. Read its mass.

2. Pour the liquid into the beaker. Read the new mass.

3. Calculate the mass of the liquid like this:

   (mass of liquid) = (mass of beaker + liquid) – (mass of beaker)

Balances normally measure mass in grams, g, or kilograms, kg.

↑ The balance measures mass.

Working safely

Hazard symbols

Hazards are the possible dangers linked to using substances or doing experiments. Hazardous substances display **hazard symbols**. The table shows some hazard symbols. It also shows how to reduce risks from each hazard.

Hazard symbol	What it means	Reduce risks from this hazard by...
	Corrosive – the substance attacks and destroys living tissue, such as skin and eyes.	• wearing eye protection • avoiding contact with the skin
	Irritant – the substance is not corrosive, but will make the skin go red or form blisters.	• wearing eye protection • avoiding contact with the skin
	Toxic – can cause death, for example, if it is swallowed or breathed in.	• wearing eye protection • wearing gloves • wearing a mask, or using the substance in a fume cupboard
	Flammable – catches fire easily.	• wearing eye protection • keeping away from flames and sparks
	Explosive – the substance may explode if it comes into contact with a flame or heat.	• wearing eye protection • keeping away from flames and sparks
	Dangerous to the environment – the substance may pollute the environment.	• taking care with disposal

Other hazards

The table does not list all the hazards of doing practical work in science. You need to follow the guidance below to work safely. Always follow your teacher's safety advice, too.

- Take care not to touch hot apparatus, even if it does not look hot.
- Take care not to break glass apparatus – leave it in a safe place on the table, where it cannot roll off.
- Support apparatus safely. For example, you might need to weigh down a clamp stand if you are hanging heavy loads from the clamp.
- If you are using an electrical circuit, switch it off before making any change to the circuit.
- Remember that wires may get hot, even with a low voltage.
- Never connect wires across the terminals of a battery.
- Do not look directly at the Sun, or at a laser beam.
- **Wear eye protection** – *whatever* you are doing in the laboratory!

Recording results

A simple table

Results are easier to understand if they are in a clear table.

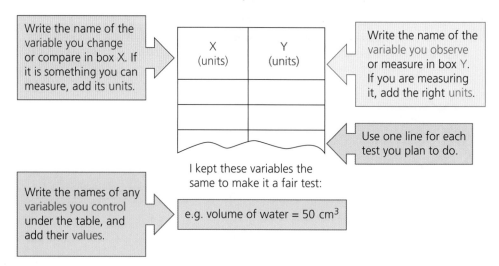

Write the name of the variable you change or compare in box X. If it is something you can measure, add its units.

Write the name of the variable you observe or measure in box Y. If you are measuring it, add the right units.

Use one line for each test you plan to do.

Write the names of any variables you control under the table, and add their values.

I kept these variables the same to make it a fair test:

e.g. volume of water = 50 cm³

Units

It is very important to use the correct units.

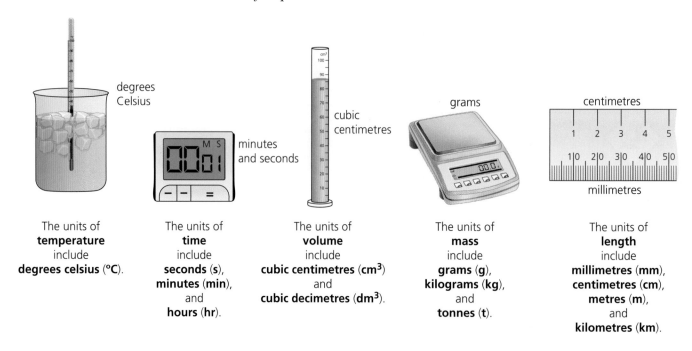

degrees Celsius

minutes and seconds

cubic centimetres

grams

centimetres

millimetres

The units of **temperature** include **degrees celsius (°C)**.

The units of **time** include **seconds (s)**, **minutes (min)**, and **hours (hr)**.

The units of **volume** include **cubic centimetres (cm³)** and **cubic decimetres (dm³)**.

The units of **mass** include **grams (g)**, **kilograms (kg)**, and **tonnes (t)**.

The units of **length** include **millimetres (mm)**, **centimetres (cm)**, **metres (m)**, and **kilometres (km)**.

Making results reliable, and reducing error

You should always try to repeat observations or measurements. Never rely on a single result. If repeat results are similar, they are more **reliable**. Repeating results also helps to reduce error. If the results keep changing, try to find out why. It could be a mistake, or there might be another variable you need to control.

When you collect similar measurements, you should calculate their average value.

Three students find the time it takes to draw a table. Jamil takes 75 seconds, Abiola takes 35 seconds, and Karis takes 73 seconds.

Abiola's result is **anomalous** because it is very different from the others. Jamil and Karis find out why. Abiola's table is very messy. She did not use a ruler. They decide to leave it out of the average.

$$\textbf{average time} = \frac{\textbf{sum of the measurements}}{\textbf{number of measurements}} = \frac{75 + 73}{2} = \frac{148}{2} = \textbf{74 seconds}$$

A table for repeat results

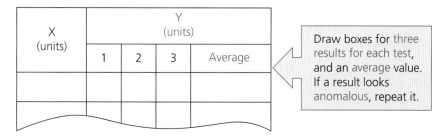

X (units)	Y (units)			
	1	2	3	Average

Draw boxes for three results for each test, and an average value. If a result looks anomalous, repeat it.

A table for results that need to be calculated

Some variables can't be measured. They need to be calculated from two different results. Do all the calculations before you calculate an average value.

X (units)	Result 1 (units)	Result 2 (units)	Y (units)	Average (units)

Draw boxes for three sets of measurements, three calculated values, and an average. If a result looks anomalous, repeat it.

Displaying results

Drawing a bar chart

Three students timed how long they spent doing homework. The results are in the table.

A **bar chart** makes results like these easier to compare.

Students	Time spent on homework (hours)
Deepak	2
Jamila	4.5
Kasim	1

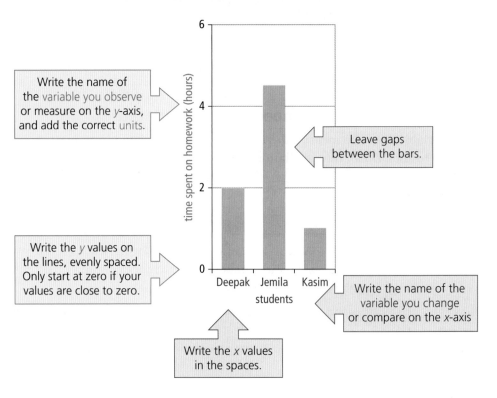

Write the name of the variable you observe or measure on the y-axis, and add the correct units.

Leave gaps between the bars.

Write the y values on the lines, evenly spaced. Only start at zero if your values are close to zero.

Write the name of the variable you change or compare on the x-axis

Write the x values in the spaces.

Categoric, discrete, or continuous

If the values of the variable you change (*x*) are *words*, then *x* is a **categoric** variable.

Variables like shoe size are **discrete** variables. They are numbers, but there are no inbetween sizes. The number of paper clips or number of people are discrete variables. You can only draw a bar chart or a **pie chart** for data that include categoric or discrete variables.

Other variables are **continuous** variables. Their values can be any number. Height is a continuous variable and so is temperature or time.

If the variables you change and measure are both continuous variables, display the results on a **line graph** or **scatter plot**.

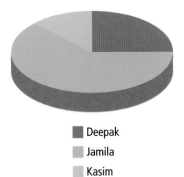

Deepak
Jamila
Kasim

Drawing a line graph

A line graph makes it easier to see the **link** between two continuous variables – the variable you change or compare and the variable you observe or measure.

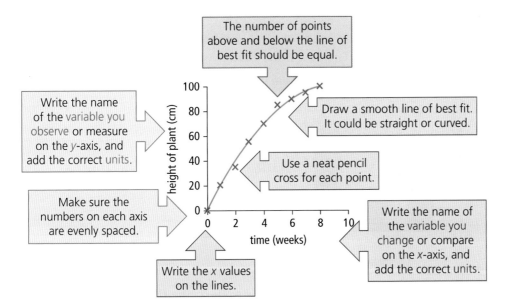

A line graph shows the link between two variables. If there is no link then changing one thing will have no effect on the other. The line graph would be a straight horizontal line.

Drawing a scatter graph

A scatter graph shows whether there is a **correlation** between two continuous variables. In the graph below, all the points lie close to a straight line. That means there is a correlation between them.

A correlation does not mean that one variable affects the other one. Something else could make them both increase or decrease at the same time. If there is no correlation between the variables then the points would be scattered all over the graph.

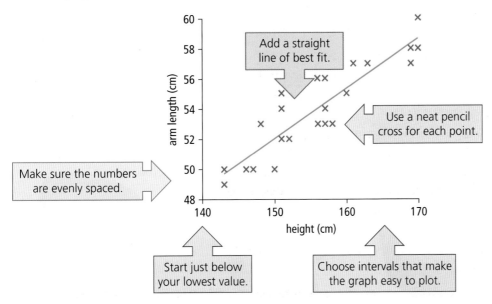

Analysing results: charts and diagrams

Charts and diagrams help you to analyse results. They show differences between results clearly. The flow chart shows you how to analyse results.

Here is a bar chart to show how long it takes three students to get to school.

Describe the pattern.
(e.g. A is bigger than B)

↓

Use the numbers to compare results.
(e.g. A is three times bigger than B)

↓

Suggest a reason using scientific knowledge.
(e.g. This is because...)

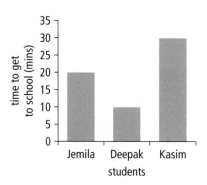

The first stage of analysing your results is to **describe the pattern**.

> "Deepak takes the shortest time to get to school. Kasim takes the longest time to get to school."

The next stage it to **use the numbers** on the *y*-axis of the bar chart to **make comparisons**.

> "Deepak takes 10 minutes to get to school, but Kasim takes 30 minutes. Kasim takes three times longer."

Finally, **suggest reasons** for any differences that you have found.

> "This could be because Kasim lives further away or because Deepak walks more quickly."

When you do an experiment, use **scientific knowledge** to **explain** differences between results. Here are some examples of scientific reasons.

I think that the powdered sugar dissolved faster than the normal sugar **because** the pieces were much smaller.

I think that the plant near the window grew more quickly than the plant away from the window **because** there was more sunlight there.

I think that the shoe on the carpet needed a bigger force to move it than the shoe on the wooden floor **because** there was more friction between the shoe and the carpet.

Analysing results – line and scatter graphs

Line or scatter graphs show correlations between continuous variables. When you have plotted the points on a line or scatter graph, draw a line of best fit. Then analyse the graph. The flow chart on the right shows how to do this.

In the graphs below the line of best fit is shown, but not the points.

Describe the pattern by saying what happens to B as A increases. (e.g. As A increases B increases)

Choose pairs of values to illustrate the pattern and compare them. (e.g. When A is 3, B is 2, and when A is 6 B is 4 so doubling A will double B)

Suggest a reason using scientific knowledge. (e.g. This is because...)

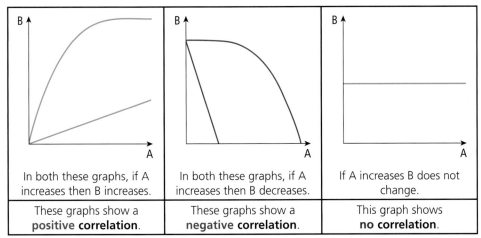

In both these graphs, if A increases then B increases.	In both these graphs, if A increases then B decreases.	If A increases B does not change.
These graphs show a **positive correlation**.	These graphs show a **negative correlation**.	This graph shows **no correlation**.

On some graphs the line of best fit is a straight line. You can say 'B changes by the same amount for each increase in A'. The blue line in graph 1 shows that B increases by the same amount for each change in A. The red line shows that B decreases by the same amount for each change in A. You can choose values to illustrate this. Graph 2 on the right shows how.

A graph with a positive correlation where the line of best fit is a straight line that *starts at zero* is a special case. In this case B is **directly proportional** to A. You can say 'if A doubles then B will also double'. (See graph 3.) You can choose values to illustrate this.

A graph that has a curved shape with a negative correlation may show that B is **inversely proportional** to A. Choose pairs of values. If you get the *same number* every time you multiply A by B then B is inversely proportional to A. This is the same as saying 'if you double A then B will halve'. (See graph 4.) You can choose values to illustrate this.

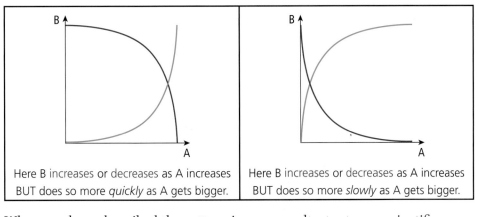

Here B increases or decreases as A increases BUT does so more *quickly* as A gets bigger.	Here B increases or decreases as A increases BUT does so more *slowly* as A gets bigger.

When you have described the pattern in your results, try to use scientific knowledge to explain the pattern.

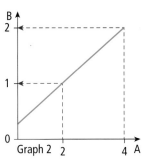

Graph 1

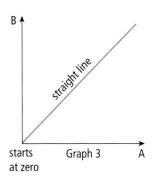

Graph 2

starts at zero   Graph 3

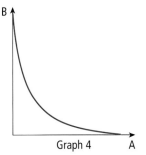

Graph 4

Using ammeters and voltmeters

Measuring current and voltage

In your experiments with electrical circuits you will measure current and voltage using meters. There are two types of meter: analogue and digital.

- Analogue meters have a needle that moves and shows the current on a scale.

- Digital meters have a display of numbers. A digital meter is usually a **multimeter**, which means that you can use it to measure current *or* voltage.

Using an analogue ammeter

An ammeter measures the current. You might want to find the current flowing through a component such as a lamp. To do this you place the ammeter *in series* with the lamp.

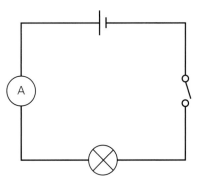

For most components in a circuit it does not matter which way round you connect them. This is not true for an ammeter. It needs to be connected so that, if you follow the wires back to the battery or cell, the wire in the *black* terminal on the ammeter is connected to the *negative* terminal of the battery. The wire in the *red* terminal should be connected to the *positive* terminal of the battery.

This is how you connect the ammeter into a circuit with a cell or battery, a lamp, and a switch.

1 Disconnect the cell or open the switch.

2 Follow the wire from the negative terminal of the cell until you get to the lamp. Disconnect the lead from the lamp.

3 Plug that lead into the black terminal on your ammeter.

4 Use another lead to connect the lamp to the red terminal of the ammeter.

An analogue meter may have two different scales. Start by connecting it up using the red terminal labelled with the higher value of current. If the ammeter does not show a current in your circuit using that scale, then move the lead to the terminal with the lower value of current.

Using an analogue voltmeter

A voltmeter measures the voltage. You might want to find the voltage across a component such as a lamp. To do this you place the voltmeter *in parallel* with the lamp.

Like an ammeter, a voltmeter needs to be connected the right way round. This is how you connect the voltmeter into a circuit with a cell or battery, a lamp, and a switch.

1 Disconnect the cell or open the switch.

2 Follow the wire from the negative terminal of the cell until you get to the lamp. Do not disconnect the lamp but plug another lead into the terminal on that side of the lamp.

3 Connect the other end of this lead to the black terminal on your voltmeter.

4 Use another lead to connect the other side of the lamp to the red terminal of the voltmeter.

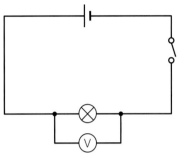

Connecting meters

You connect an ammeter in series, and a voltmeter in parallel. It is important not to connect an ammeter in parallel with a component, or directly across a battery or cell. The resistance of the ammeter is very small so a very large current would flow that could damage the ammeter and drain the cell.

The resistance of a voltmeter is very, very high. If you connect it in series rather than in parallel no current will flow in the circuit.

Using a multimeter

You can use a multimeter to measure current *or* voltage.

The multimeter has a dial that you turn to select current (A or mA) or voltage (V). You select whether you want to measure a large current (10 A) or a small current. Some examples of how to connect a multimeter are shown below.

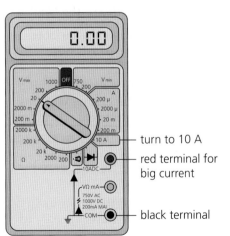

▲ Using a multimeter to measure a big current.

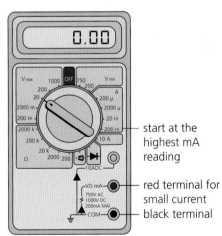

▲ Using a multimeter to measure a small current.

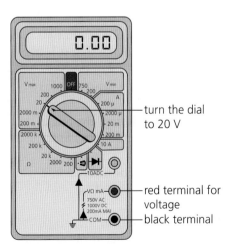

▲ Using a multimeter to measure voltage.

Glossary

Absorbed energy of electromagnetic radiation (e.g. light) or sound is transferred to thermal energy on passing through or into a medium

Accelerate speed up, get faster

Acceleration the rate of change of increasing speed (the amount by which speed increases in one second)

Accuracy (of a measurement) how correct a measurement is – how close to its true value

Air resistance the force on an object that is moving through the air, causing it to slow down (also known as drag)

Alloy a material made of a mixture of metals, or of carbon with a metal

Ammeter a device for measuring electric current in a circuit

Ampere (amp) the unit of measurement of electric current, symbol A

Amplifier a device for making a sound louder

Amplitude the distance from the middle to the top or bottom of a wave

Analogy a way of explaining something by saying that it is like something else

Andromeda the nearest galaxy to the Milky Way

Angle of incidence the angle between the incident ray and the normal line

Angle of reflection the angle between the reflected ray and the normal line

Angle of refraction the angle between the refracted ray and the normal line

Anomalous point a point on a graph that does not fit the general pattern (also called an outlier)

Anticlockwise the direction of rotation that is opposite to the movement of the hands of a clock

Apparent depth how deep something underwater appears to be when viewed from above

Archimedes' principle a law that says the upthrust on an object is equal to the weight of water displaced

Armature the coil of wire in an electromagnetic device such as a generator

Artificial satellites spacecraft made by people to orbit the Earth for various purposes

Asteroid a lump of rock in orbit around the Sun

Asteroid belt a large number of asteroids between Mars and Jupiter

Astronomer a scientist who studies space

Atmosphere the layer of air above Earth's surface

Atmospheric pressure the force on an area of the Earth's surface due to the weight of air above it, or the pressure in the atmosphere

Atom the smallest particle of an element that can exist

Attract pull together, for example opposite poles of a magnet or positive and negative charges attract each other

Audible can be heard

Auditory canal the passage in the ear from the outer ear to the eardrum

Auditory nerve signals are sent from your ear to the brain along this nerve

Average found by adding a set of values together and dividing by the number of values (also called the mean)

Average speed the total distance travelled divided by the total time taken for a complete journey

Average speed (molecules) the typical value of the speed of molecules in a gas or liquid

Axis, of Earth the imaginary line through the Earth around which it spins

Balanced forces describes forces that are the same size but act in opposite directions on an object

Bar chart a way of presenting data in which only one variable is a number

Battery two or more electrical cells joined together

Big Bang the expansion of space which we believe started the Universe

Biodiesel a biofuel made from plant oils

Bioethanol a biofuel made from carbohydrates such as sugar

Biofuel a fuel produced from renewable resources

Biogas gas produced from waste products, usually methane, used to generate electricity

Biomass material from plants used for fuel, e.g. wood

Black dwarf a remnant of a star like our Sun that no longer gives out light

Black hole a remnant of a star much bigger than our Sun from which nothing can escape, not even light

Boil to change from a liquid into a gas at the boiling point

Capacitor two plates with an insulator between them, used to store charge

Carbon dioxide a gas found in small amounts in the atmosphere which plants use to make food and which is a greenhouse gas

Categoric describes a variable whose values are words not numbers

Cell (electrical) a device that is a store of chemical energy which is transferred to electrical energy in a circuit

Centre of gravity the point in an object where the force of gravity acts

Centre of mass the point in an object where the mass appears to be concentrated

Centripetal force the force directed towards the centre that causes a body to move in a uniform circular path

Charge positive or negative, a property of protons and electrons

Charge-coupled device (CCD) a grid of components that work like capacitors at the back of a digital camera, and produce a digital signal

Chemical potential energy (chemical energy) energy stored in fuels, food, and electrical batteries

Circuit (electric) a complete pathway for an electric current to flow

Circuit diagram a way of showing a circuit clearly, using symbols

Circuit symbol a drawing that represents a component in a circuit

Climate change changes to long-term weather patterns as a result of global warming

Clockwise the direction of rotation that is the same as the movement of the hands of a clock

Coal a fossil fuel formed from dead plants that have been buried underground over millions of years

Coal-fired power station a place where the fossil fuel coal is burned to generate electricity

Cobalt a magnetic material

Cochlea a snail-shaped tube in the inner ear where the sensory cells that detect sound are

Colour blindness someone with colour blindness cannot tell certain colours apart, because some cone cells in the retina of the eye do not work properly

Comets bodies in space made of dust particles frozen in ice, which orbit the Sun

Communicate to share and exchange information

Compact disc a metal disc that can store high-quality digital recordings

Compass a device containing a small magnet that is used for finding directions

Component an item used in an electric circuit, such as a lamp

Compressed squashed into a smaller space

Compression the part of a sound wave where the air particles are close together

Conclude to make a statement about what the results of an investigation tell you

Conduction (of charge) the movement of charge through a material such as metal or graphite, forming an electric current

Conduction (of thermal energy) the way in which thermal energy is transferred through solids (and to a much lesser extent in liquids and gases)

Conductor a material such as a metal or graphite that conducts charge or thermal energy well

Cone a specialised cell in the retina that is sensitive to bright light and colour

Consequence (risk) what can happen as a result of something you do

Conservation (of energy) a law that says that energy is never made or lost but is always transferred from one form to another

Constant not changing

Constellation a collection of stars that make a pattern in the sky

Contact force a force that acts when an object is in contact with a surface, air, or water

Continuous variable a variable that can have any value across a range, such as time, temperature, length

Convection the transfer of thermal energy by the movement of a gas or liquid

Convection current the way in which thermal energy is transferred through liquids and gases by the movement of their particles

Conventional current the current that flows from the positive to the negative terminal of a battery

Core a rod of a magnetic material placed inside a solenoid to make the magnetic field of an electromagnet stronger

Cornea the transparent layer at the front of the eye

Correlation a link between two things; it does not necessarily mean that one thing causes the other

Creative thinking thinking in a new way

Crescent moon the shape of moon that we see when a thin section of the Moon is lit, a few days after a new moon

Critical angle the smallest angle of incidence at which total internal reflection occurs

Crude oil a thick black liquid formed underground from the remains of prehistoric plants and animals that died millions of years ago. It is used to make fuels such as petrol/gasoline and diesel, and many plastics.

Current the flow of electric charge (electrons) around a complete circuit

Data measurements taken from an investigation or experiment

Day the period of time when one section of the Earth (or other planet) is facing the Sun

Decelerate slow down, get slower

Deceleration the amount by which speed decreases in one second

Decibel (dB) a commonly used unit of sound intensity or loudness

Deform change shape

Degrees Celsius (°C) a temperature scale with 0 °C fixed at the melting point of ice and 100 °C fixed at the boiling point of water

Demagnetise to destroy a magnet by heating it up, hitting it, or putting it in an alternating current

Density the mass of a substance in a certain volume

Dependent variable the variable that changes when you change the independent variable

Detector something that absorbs electromagnetic radiation or sound to produce a signal

Dielectric the insulating material between the plates of a capacitor

Diffuse particles in gases and liquids move from where there are a lot of the particles to where there are fewer

Diffuse describes reflection from a rough surface

Diffusion the way particles in liquids and gases mix or spread out by themselves

Digital signal a signal that is used to transfer information between sensors and computers

Directly proportional a relationship in which one quantity increases in the same way as another

Discrete describes a variable that can only have whole-number values

Dispersion the splitting up of a ray of light of mixed wavelengths by refraction into its components

Distance multiplier a type of lever that uses a larger force to produce a smaller force at a larger distance

Distance–time graph a graph showing how the distance travelled varies with time

Domain a small region inside a magnetic material that behaves like a tiny magnet

Drag a force on an object moving through air or water, causing it to slow down

Dwarf planet a lump of rock in orbit around the Sun that is nearly spherical but has other objects around it

Dynamo a device that transforms kinetic energy into electrical energy (a small generator)

Ear defenders a device used to protect the ears from noise

Eardrum a membrane that transmits sound vibrations from the outer ear to the middle ear

Earth a rocky inner planet, the third planet from the Sun

Earth (charge) to connect a metal wire from an object to the ground to take any charge away

Earthing the process of connecting objects to the ground

Echo a reflection of a sound wave by an object

Echolocation the process of finding out where something is using echoes

Eclipse the Sun or Moon is blocked from view on Earth (see also lunar eclipse or solar eclipse)

Efficient describes something that does not waste much energy

Effort the amount of force that you use to push down when using a lever

Elastic describes a type of material that can be stretched and will return to its original length when the pulling force is removed

Elastic limit the point beyond which a spring will never return to its original length when the pulling force is removed

Elastic potential energy (EPE) energy stored in an elastic object that is stretched or squashed

Electric car car powered by electric batteries

Electric circuit a complete pathway for an electric current to flow

Electric current a flow of electric charge (electrons) around a complete circuit

Electric field a region around a charged object where other charged objects feel a force

Electrical energy energy that is transferred from a cell or battery to the components of a circuit. A light bulb changes electrical energy to light and thermal energy.

Electrical signal (ear) information is transferred from the ear to the brain as an electrical signal (nerve signal)

Electromagnet a temporary magnet produced using an electric current

Electromagnetic spectrum range of radiation with electrical and magnetic properties that can travel through a vacuum (such as the Sun's radiation)

Electromagnetic spectrum the range of wavelengths of electromagnetic radiation produced by the Sun and other sources

Electron tiny charged particle in an atom, that flows through a wire to create an electric current

Electrostatic force the force between two charged objects

Electrostatic phenomena things that happen because objects have become charged

Element a substance consisting of atoms of only one type

Emit to send something out (such as heat, light, vapour)

Endoscope a medical instrument for seeing inside the human body

Energy this is needed to make things happen

Energy conservation energy is never made or lost but may be transferred from one form to another, although they are not always forms we can use

Energy converter something that produces a secondary source of energy, such as a power station

Energy transfer energy changing from one form to another, such as from chemical to thermal energy

Energy transfer diagram a diagram that shows how energy is transferred or changed in a process or device

Equator an imaginary line round the middle of the Earth at an equal distance from both the North and South Poles

Equilibrium balanced (as in a lever or see-saw)

Evaporate to turn from a liquid to a vapour (gas)

Evaporation the process of turning from a liquid to a gas without boiling

Evaporative coolers cooling equipment used in hot countries that use the cooling effect of evaporation to cool houses

Evidence observations and measurements that support or disprove a scientific theory

Exojoule a million, million, million joules

Exoplanet a planet in orbit around a star other than our Sun

Expand to increase in size, get bigger

Explanation A statement that gives a reason for something using scientific knowledge

Extension the amount an object gets longer when you stretch it

Eye the organ of sight, which focuses and detects light

Far side of the moon the side of the Moon that we never see because the Moon is spinning on its axis, sometimes called the 'dark side of the Moon'

Field study an experiment or observations made in the natural habitat of an organism

Filament the very thin, coiled piece of wire inside a light bulb that glows

Filter a piece of material that allows some radiation (colours) through but absorbs the rest

First quarter the shape of moon that we see when about half the Moon is lit, about a week after a new moon

Floating an object floats when the upthrust from water is equal to the downwards force of the object's weight

Force a push or a pull that acts on an object to affect its movement or shape

Force multiplier a lever or hydraulic machine that can lift or move heavy weights using a force smaller than the weight

Forcemeter a device used to measure forces

Fossil fuel fuels made from the decayed remains of animals and plants that died millions of years ago. Fossil fuels include coal, oil, and natural gas.

Frequency the number of complete waves or vibrations produced in one second (measured in hertz)

Friction a force that resists movement because of contact between surfaces

Fuel a material that contains a store of energy and can be burned, e.g. gas, oil, coal, petrol (gas or gasoline)

Fuel cell a device that uses chemical reactions to generate electricity

Fulcrum the point about which a lever or see-saw turns

Full moon the shape of moon that we see when the whole disc is lit, when the Moon is opposite the Sun

Fundamental (sound) the lowest frequency of sound

Galaxy a number of stars and the solar systems around them grouped together

Gas (natural) a fossil fuel that collects above oil deposits underground

Gas pressure (air pressure) the force exerted by air particles when they collide with 1 square metre (1 m²) of a surface

Gemstone a stone made from minerals that can be cut and used in jewellery

Generator a device that uses kinetic energy to induce a voltage

Geocentric model a model of the Universe with the Earth at the centre

Geothermal an energy source that uses the thermal energy underground to produce electricity

Gibbous moon the shape of the moon that we see when most of the Moon is lit, a few days after a full moon

Global positioning system (GPS) a system that pinpoints the position of something using signals from a satellite

Gradient the slope or steepness of a graph

Gravitational field a region in which there is a force on a mass due to its attraction to another mass

Gravitational field strength the force on a mass of 1 kg, measured in N/kg

Gravitational force (gravity) the force of attraction between two objects because of their mass

Gravitational potential energy (GPE) energy stored in an object because of its height above the ground

Greenhouse effect the effect of heating the Earth because the energy from the Sun is reflected back to the Earth from a layer of greenhouse gases in the atmosphere

Greenhouse gases gases that contribute to global warming, such as carbon dioxide, water vapour, and methane

Harmonics frequencies of a sound wave that are multiples of the fundamental frequency

Hazard symbols warning symbols used on chemicals which show what harm they might cause if not handled properly

Heat to change the temperature of something; the word 'heat' is sometimes used instead of thermal energy

Heat pump a device that transfers heat from the ground to a building on the surface

Heliocentric model a model of the Universe with the Sun at the centre

Hertz (Hz) the unit of frequency

Hubble space telescope a telescope orbiting the Earth and sending back clear pictures of space

Hybrid car a car that can run either on electricity from a battery or on petrol/gasoline in a petrol engine

Hydraulic brakes brakes that use a liquid in pipes to transfer forces and make them bigger

Hydraulic machine a machine that uses a liquid in pipes to transfer forces and make them bigger

Hydraulic press a press that uses a liquid in pipes to transfer forces and make them bigger

Hydroelectricity electricity generated using the energy of water falling downhill

Hydrogen a non-metal element that exists as a gas at everyday temperatures

Hydrogen cars cars that are designed to use hydrogen from fuel cells to power the engine

Image the point from which rays of light entering the eye appear to have originated

Incident ray the ray coming from a source of light

Incompressible describes something that cannot be compressed (squashed)

Independent variable the variable that you change, that causes changes in the dependent variable

Induced (voltage) a voltage produced when a conductor is in a changing magnetic field

Inertia the tendency of an object to resist a change in speed caused by a force

Infinite without end

Infrared (radiation) a type of electromagnetic radiation that transfers thermal energy from a hotter to a colder place, also known as heat

Inner ear the part of the ear made up of the cochlea and semi-circular canals

Inner planets Mercury, Venus, Earth, and Mars

Insulator a material that does not conduct thermal energy or electricity very well

Intensity (sound) how loud a sound is, measured in decibels

International Space Station (ISS) a research station in orbit around the Earth

Interstellar space the space between stars or solar systems

Inversely proportional a relationship in which one quantity decreases as the other increases

Inverted upside down

Investigation an activity such as an experiment or set of experiments designed to produce data to answer a scientific question or test a theory

Iris the coloured part of your eye that controls the size of the pupil

Iron a metal element that is the main substance in steel. Iron is a magnetic material.

Iron core an iron rod placed in a coil to increase the magnetic field strength when a current flows in the coil

Joule the unit of energy, symbol J

Jupiter a large outer planet made of gas, fifth from the Sun

Kaleidoscope a toy containing mirrors and coloured glass or paper

Kilogram the unit of mass, symbol kg

Kilojoule (kJ) 1000 joules

Kilometre per hour the unit of speed, km/h

Kilowatt 1000 watts

Kinetic energy movement energy

Kuiper belt the region outside the Solar System where astronomers think that some comets come from

Laser a device that produces an intense beam of light that does not spread out

Last quarter the shape of moon that we see when about half the Moon is lit, about a week after a full moon

Laterally inverted the type of reversal that occurs with an image formed by a plane mirror

Law of conservation of energy the law that says that energy cannot be created or destroyed

Law of reflection the law that says that the angle of incidence is equal to the angle of reflection

Lens a device made of shaped glass which focuses light rays from objects to form an image

Lever a simple machine consisting of a rigid bar supported at a point along its length

Life cycle (of a star) the process that describes how a star is formed and what will happen to it

Light a form of electromagnetic radiation that comes from sources like the Sun

Light-emitting diode (LED) a low-energy lamp

Light energy energy transferred by sources such as the Sun and light bulbs

Light intensity the energy per square metre, measured in lux

Light sources objects that emit visible light, also called luminous objects

Light year the distance light travels in one year

Lightning conductor a piece of metal connected to tall buildings to conduct lightning to the ground

Line graph a way of presenting results when there are two numerical variables

Line of best fit a smooth line on a graph that travels through or very close to as many of the points plotted as possible

Liquid pressure the pressure produced by collisions of particles in a liquid

Load an external force that acts over a region of length, surface, or area

Lodestone a naturally occurring magnetic rock

Longitudinal describes a wave in which the vibrations are in the same direction as the direction the wave moves

Loudspeaker a device that changes an electrical signal into a sound wave

Lubrication reducing friction between surfaces when they rub together

Luminous describes something that gives out light

Lunar eclipse occurs when the light from the Sun is blocked by the Earth because it is in a direct line between the Earth and the Sun, and the Moon is in shadow

Lux the unit of light intensity

Mach the ratio of the relative speed to the speed of sound

Magnet an object that attracts magnetic materials and repels other magnets

Magnetic field an area around a magnet where there is a force on a magnetic material or another magnet

Magnetic field lines imaginary lines that show the direction of the force on a magnetic material in the magnetic field

Magnetic force the force between the poles of two magnets, or between a magnet and a magnetic material such as iron

Magnetic material a material that is attracted to a magnet, such as iron, steel, nickel, or cobalt

Magnetic resonance imaging (MRI) scanner a machine that uses strong magnetic fields to produce images of the inside of the human body

Magnetised made into a magnet

Magnetism the property of attracting or repelling magnets or magnetic materials

Main sequence the longest stage of a star's life cycle; the current stage of our Sun

Mains electricity (mains supply) electricity generated in power stations and available through power sockets in buildings

Mars a rocky planet, fourth from the Sun

Mass the amount of matter in an object. The mass affects the acceleration for a particular force.

Massive (star) having lots of mass (massive does not mean large)

Measuring cylinder a cylinder used to measure the volume of a liquid

Medium (sound/light) the material that affects light or sound by slowing it down or transferring the wave

Meniscus the curved upper surface of a liquid

Mercury the rocky inner planet nearest the Sun

Meteor a piece of rock or dust that makes a streak of light in the night sky

Meteorite a stony or metallic object that has fallen to Earth from outer space without burning up

Metre per second the unit of speed, m/s

Microphone a device for converting sound into an electrical signal

Middle ear the eardrum and ossicles (small bones) that transfer vibrations from the outer ear to the inner ear

Milky Way the galaxy containing our Sun and Solar System

Milliamp one thousandth of an amp

Minerals chemicals in rocks

Model a simplified description of a process. A model may be a physical model built on a different scale to the original system, or it may take the form of equations.

Moment a measure of the ability of a force to rotate an object about a pivot

Monochromatic describes light of a single colour or wavelength

Moon a rocky body orbiting Earth; it is Earth's only natural satellite

Moons the natural satellites of planets

Movement energy the energy of movement, also called kinetic energy

Natural satellite a moon in orbit around a planet

Nebula a region of dust and gas where stars are born

Negative describes the charge on an electron, or the charge on an object that has had electrons transferred to it

Negatively charged describes an object that has had electrons transferred to it

Neptune a large outer planet made of gas, eighth from the Sun

Neutral describes an object that has no charge; its positive and negative charges cancel out

Neutral point (magnetic field) a point where there is no force on a magnet or magnetic material because two or more magnetic fields cancel out

Neutralise to cancel out, when you add an equal amount of positive charge to negative charge

Neutron star a very small, massive star

New moon the shape of moon that we see when the Moon is in between the Earth and the Sun

Newton the unit of force including weight, symbol N

Newtonmetre the unit of moment, symbol Nm

Nickel a magnetic material

Night the period on one section of the Earth or other planet when it is facing away from the Sun

Noise any undesired or unwanted sound

Non-contact force a magnetic, electrostatic, or gravitational force that acts without being in contact with something

Non-luminous describes objects that produce no light; objects that are seen by reflected light

Non-renewable describes an energy source that will run out eventually (such as fossil fuels)

Normal an imaginary line at right angles to a surface where a light ray strikes it

Normal brightness standard brightness of a single bulb lit by a single cell

North pole the pole of a magnet that points north. A north pole repels another north pole.

Northern hemisphere the half of the Earth between the equator and the North Pole

Nuclear energy the energy from nuclear fusion that powers the Sun and stars, or from uranium in nuclear power stations

Nuclear fusion the process of joining hydrogen together in the Sun and other stars that releases energy

Nucleus the centre of an atom that contains neutrons and protons

Object something that can be seen or touched

Observations the results of looking carefully at something and noticing properties or changes

Oil a fossil fuel formed from sea creatures over millions of years

Oort cloud a cloud of comets and dust outside the Solar System

Opaque describes objects that absorb, scatter, or reflect light and do not allow any light to pass through

Optic nerve a sensory nerve that runs from the eye to the brain

Optical fibre a very fine tube of plastic or glass that uses total internal reflection to transmit light

Orbit the path taken by one body in space around another (such as the Earth around the Sun)

Oscilloscope a device that enables you to see electrical signals that change, such as those made in a microphone

Ossicles the small bones of the middle ear (hammer, anvil, and stirrup) that transfer vibrations from the eardrum to the oval window

Outer ear the pinna and auditory canal

Outer planets Jupiter, Saturn, Uranus, and Neptune

Oval window the membrane that connects the ossicles to the cochlea in the ear

Oxide a compound made when an element combines with oxygen

Parallel circuit an electric circuit in which there are two or more paths for an electric current

Partial eclipse occurs when part of the light from the Sun is blocked by the Moon or the Earth

Particles tiny pieces of matter

Pascal the unit of pressure, symbol Pa, equal to 1 N/m²

Payback time the time taken to recoup the cost of something, such as home insulation or a wind turbine

Pendulum any rigid body that swings about a fixed point

Penumbra the area of blurred or fuzzy shadow around the edges of the umbra

Perforate to make a hole in something

Period the time taken to complete one cycle of motion

Periscope a tube with mirrors or prisms that enables you to see over objects

Permanent magnet a piece of metal that stays magnetic

Permanently extended the irreversible extension of a spring when loaded beyond its elastic limit

Petrol (gas or gasoline) a hydrocarbon fuel (containing hydrogen and carbon) that comes from crude oil

Phases of the Moon parts of the Moon we see as it orbits the Earth

Photosynthesis the process by which green plants make their own food from carbon dioxide and water using solar energy

Pie chart a way of presenting data in which only one variable is a number

Pinna the outside part of the ear that we can see

Pitch a property of sound determined by its frequency

Pivot a support on which a lever turns or oscillates

Pixel the small square that digital images are made of; a picture element

Plane mirror a mirror with a flat reflective surface

Planet any large body that orbits a star in a solar system

Plastic a type of material that can be stretched and does not return to its original length

Pluto used to be regarded as the ninth and last planet from the Sun; now called a dwarf planet together with others of the same size that are beyond Pluto's orbit

Poles, of Earth the north and south points of the Earth connected by its axis of tilt

Poles, of magnet the opposite and most strongly attractive parts of a magnet

Positive (charge) describes the charge on a proton, or the charge on an object that has had electrons transferred away from it

Positively charged describes an object that has had electrons transferred away from it

Potential energy stored energy

Power the rate of transfer of energy, measured in watts

Power station a place where fuel is burned to produce electricity

Precision the number of decimal places given for a measurement

Prediction statement saying what you think will happen

Preliminary work the work that you do before or during the planning stage of an investigation, to work out how to do it

Pressure the force applied by an object or fluid divided by the area of surface over which it acts

Pressure gauge an instrument for measuring pressure in a liquid or gas

Primary colours of light are red, blue, and green

Primary data data collected directly by scientists during a particular investigation

Primary source sources of data that you have collected yourself

Primary sources (of energy) energy sources from the environment or underground, such as coal, uranium, or the wind

Principle of moments law that says that the clockwise moments equal the anticlockwise moments

Prism a triangular-shaped piece of glass used to produce a spectrum of light

Probability (risk) the chance that something will happen

Property a characteristic, for example wavelength and amplitude are properties of a wave

Proportional a relationship in which two variables increase at the same rate, for example when one is doubled the other doubles too

Proxima Centauri the nearest star to our Sun

Pupil the hole in the front of your eye where the light goes in

Question in science, a problem that is stated in a form that you can investigate

Rainbow an optical phenomenon that appears as the colours of the spectrum when falling water droplets are illuminated by sunlight

Rarefaction the part of a sound wave where the air particles are most spread out

Ray diagram a model of what happens to light shown by drawing selecting rays

Reaction time in humans, the time the brain takes to process information and act in response to it

Real describes an image that you can put on a screen, or the image formed in your eyes

Real depth the depth underwater that an object actually is

Receiver (sonar) a device that absorbs sound waves

Red giant part of the life cycle of a star like our Sun when it becomes much bigger and cooler

Red supergiant the next stage in the life cycle of our Sun

Reed switch a switch that uses a magnet to work

Refinery a place where crude oil is refined and separated into fuels

Reflected ray the ray that is reflected from a surface

Reflection the change in direction of a light ray or sound wave after it hits a surface and bounces off

Refract to change direction because of a change in speed

Refraction the change in direction of a light ray as a result of its change in speed

Refractive index a measure of how much light slows down in a medium

Refrigerant a liquid used in a refrigerator

Refrigerator a machine for keeping things cold using evaporation

Relay electrical device that allows current flowing through it in one circuit to switch on and off a larger current in a second circuit

Reliable describes an investigation in which very similar data would be collected if it was repeated under the same conditions

Renewable describes energy resources that are constantly being replaced and are not used up, such as falling water or wind power

Repel to push away

Reservoir a large amount of water behind a dam used in hydroelectric power

Resistance how difficult it is for current to flow through a component in a circuit

Resultant force the single force equivalent to two or more forces acting on an object

Retina the layer of light-sensitive cells at the back of the eye

Reverberation the persistence of a sound for a longer period than normal

Risk a combination of the probability that something will happen and the consequence if it did

Rod a specialised cell in the retina that is sensitive to dim light

Sankey diagram a diagram that shows all the energy transfers taking place in a process, and the amount of energy in each transfer

Satellite any body that orbits another (such as the Moon or a weather satellite around Earth)

Saturn a large outer planet made of gas, sixth from the Sun

Scatter plot a graph that shows all the values in a set of measurements

Seasons changes in the climate during the year as the Earth moves around its orbit

Secondary colours colours that can be obtained by mixing two primary colours

Secondary data data collected by other scientists and published

Secondary source sources of data collected by others that you use

Secondary sources (of energy) sources of energy that are produced from primary sources, such as electricity produced from coal, or petrol/gasoline produced from crude oil

Semicircular canals the part of the ear that helps you to balance

Series circuit an electrical circuit in which the components are joined in a single loop

Shadow an area of darkness on a surface produced when an opaque object blocks out light

Shield to put something in between a source and a receiver, for example sound is shielded by ear defenders

Signal (electrical) a voltage that changes over time

Solar cells devices that change light energy into electrical energy

Solar eclipse occurs when the Moon blocks out the light from the Sun because it is in a direct line between the Earth and the Sun, and part of the Earth is in shadow

Solar energy energy from the Sun which can be used directly to heat water, or to make electricity

Solar panels devices that transform light energy from the Sun into thermal energy

Solar System the Sun (our star) and the planets and other bodies in orbit around it. There are other solar systems in the Universe as well as our own.

Solenoid a core of wire used to make an electromagnet

Sonar a system that uses ultrasound to detect underwater objects or to determine the depth of the water

Sound vibrations of molecules produced by vibrating objects and detected by your ear

Sound energy energy produced by vibrating objects

Sound level meter a device for measuring the intensity (loudness) of a sound

Sound wave a series of compressions and rarefactions that moves through a medium

Source (of light/sound) something that emits (gives out) light or sound

South pole the pole of a magnet that points south. A south pole attracts a north pole.

Southern hemisphere the half of the Earth between the equator and the South Pole

Spark a flash of light that you see when the air conducts electricity

Spectrum a band of colours produced when light is spread out by a prism

Speed the distance travelled in a given time, usually measured in metres per second, m/s

Speed of light the distance light travels in one second (300 million m/s)

Speed–time graph a graph that shows how the speed of an object varies with time

Spring balance a device for measuring forces, sometimes called a forcemeter or a newton meter

Spring a metal wire wound into spirals that can store elastic potential energy

Stable describes an object in equilibrium that cannot be easily toppled

Static(of charge) charge on an insulator that does not move

Star a body in space that gives out its own light. The Sun is a star.

Steady speed a speed that doesn't change

Steel an alloy of iron with carbon and other elements. Steel is a magnetic material.

Streamlining designing a shape designed to reduce resistance to motion from air or liquid

Stretch the extension when an elastic material such as a spring is pulled outwards or downwards

Sun star at the centre of our Solar System

Sunlight light from the Sun

Sunspots dark spots on the surface of the Sun

Supernova an exploding star

Supersonic describes a speed that is faster than the speed of sound

Symbol a sign that represents something (see also circuit symbols, and hazard symbols)

Tangent a straight line that touches a curve or circle

Telescope a device made with lenses that allows distant objects to be seen clearly

Temperature a measure of how hot something is

Tension a stretching force

Terminal (of a cell) the positive or negative end of a cell or battery

Terminal velocity the highest velocity an object reaches when moving through a gas or a liquid; it happens when the drag force equals the forward or gravitational force

The bends the sickness that divers can suffer due to dissolved gases in their blood

Thermal energy the energy due to the motion of particles in a solid, liquid, or gas

Thermal image an image made using thermal or infrared radiation

thermal imaging camera a device that forms an image using thermal or infrared radiation so that different temperatures appear as different colours

Thermal a rising current of heated air

Thermometer a device used to measure temperature

Thrust force from an engine or rocket

Tidal energy/power energy from the movement of water in tides which can be used to generate electricity

Timbre the quality of a sound resulting from the harmonics present in the sound

Timing gates two sensors connected together used to measure speed or acceleration precisely and accurately

Total eclipse occurs when all of the light from the Sun is blocked out by the Earth or the Moon

Total internal reflection the complete reflection of light at a boundary between two media

Transformation (of energy) a change from one type to another

Transmitter a device that gives out a signal, such as sound in a sonar transmitter

Transducer a device that changes an electrical signal into light or sound, or changes light or sound into an electrical signal

Transfer (of energy) shifting energy from one place to another

Translucent describes objects that transmit light but diffuse (scatter) the light as it passes through

Transmitted light or other radiation passed through an object

Transparent describes objects that transmit light; you can see through transparent objects

Transverse describes a wave in which the vibrations are at right angles to the direction the wave moves

Turbine a component in a generator that turns when kinetic energy is transferred to it from steam, water, or wind

Turning effect a force causing an object to turn

Turning force the moment of a force

Ultrasound sound at a frequency greater than 20 000 Hz, beyond the range of human hearing

Umbra the area of total shadow behind an opaque object where no light has reached

Unbalanced forces describes forces on an object that are unequal

Universe everything that exists

Upright describes an image that is the right way up

Upthrust the force on an object in a liquid or gas that pushes it up

Uranium a metal used in nuclear power stations

Uranus a large outer planet made of gas, seventh from the Sun

Useful energy the energy that you want from a process

Vacuum a space in which there is no matter

Variable a quantity that can change, such as time, temperature, length, or mass. In an investigation you should change only one variable at a time to see what its effect is.

Venus a rocky inner planet, second from the Sun

Vibrate to move continuously and rapidly to and fro

Vibration motion to and fro of the parts of a liquid or solid

Virtual describes an image that cannot be focused onto a screen

Volt the unit of measurement of voltage, symbol V

Voltage a measure of the strength of a cell or battery used to send a current around a circuit, measured in volts

Voltmeter a device for measuring voltage

Volume (of space) the amount of space that something takes up, which is related to its mass by its density

Waning describes the moon we see when the amount of the lit side is decreasing

Wasted energy energy transferred to non-useful forms, often thermal energy transferred to the surroundings

Water (energy from) using water to generate electricity from tides, water behind dams, and waves

Water resistance the force on an object moving through water that causes it to slow down (also known as drag)

Watts the unit of power, symbol W

Wave a variation that transfers energy

Wave energy/power using energy in waves to generate electricity

Wavelength the distance between two identical points on the wave, such as two adjacent peaks or two adjacent troughs

Waxing describes the moon we see when the amount of the lit side is increasing

Weight the force of the Earth on an object due to its mass

White dwarf a small, very dense star, part of the life cycle of our Sun

Wind energy/power energy from wind that can be used to generate electricity

Wind farm a collection of wind turbines

Wind turbine a turbine and generator that uses the kinetic energy of the wind to generate electricity

Year the length of time it takes for a planet to orbit the Sun

Index

OXFORD
UNIVERSITY PRESS

Great Clarendon Street, Oxford OX2 6DP

Oxford University Press is a department of the University of Oxford.
It furthers the University's objective of excellence in research, scholarship,
and education by publishing worldwide in

Oxford New York

Auckland Cape Town Dar es Salaam Hong Kong Karachi
Kuala Lumpur Madrid Melbourne Mexico City Nairobi
New Delhi Shanghai Taipei Toronto

With offices in

Argentina Austria Brazil Chile Czech Republic France Greece
Guatemala Hungary Italy Japan Poland Portugal Singapore
South Korea Switzerland Thailand Turkey Ukraine Vietnam

British Library Cataloguing in Publication Data

Data available

ISBN 978-0-19-839024-4

10 9 8

Printed in China

Acknowledgments

®IGCSE is the registered trademark of Cambridge International Examinations.

Cambridge International Examinations bears no responsibility for the example answers
to questions taken from its past question papers which are contained in this publication.

Cover image: PASIEKA/Getty; **p8:** Lightpoet/Dreamstime; **p8:** Tyler Panian/Fotolia;
p8: Neven M/Fotolia; **p8:** GIPhotoStock/Photo Researchers/Getty Images; **p8:** Tom
Mounsey/iStockphoto; **p9:** Melissa Madia/Shutterstock; **p11:** Gary Blakeley/Bigstock;
p12: Denise Kappa/Shutterstock; **p12:** Jeannette Meier Kamer/Shutterstock; **p13:**
Christian Mueller/Shutterstock; **p13:** Kevin Norris/Shutterstock; **p13:** rawcaptured/
Shutterstock; **p14:** Amy Walters/Pantherstock; **p15:** © NASA Images/Alamy; **p16:** paul
prescott/Shutterstock; **p17:** MarcelClemens/Shutterstock; **p17:** Huebi/Shutterstock; **p18:**
Joggie Botma /Shutterstock; **p18:** Andy Sacks/Stone/Getty Images; **p18:** Action Sports/
Bigstock; **p19:** Jorgen Udvang / Alamy; **p20:** Willyam Bradberry/shutterstock; **p22:** Greg
Epperson/Fotolia; **p22:** Vitalii Nesterchuk/Shutterstock; **p22:** © Adrianne Yzerman /
Alamy; **p23:** Gerard Koudenburg/Shutterstock; **p24:** John Lumb/Shutterstock; **p8:** Tyler
Panian/Fotolia; **p26:** William West/AFP/Getty Images; **p27:** Lexa Arts/Fotolia; **p30:** Brian
Jackson/Fotolia; **p30:** Phase4Photography/Bigstock; **p34:** Adam Gault/OJO Images/OUP
Picturebank; **p35:** Nikolai Sorokin/Fotolia; **p35:** Chris King/OUP/OUP Picturebank; **p35:**
Tryfonov/Fotolia; **p36:** Q2A Photobank; **p36:** IAN BODDY/SCIENCE PHOTO LIBRARY/Glow
Images; **p39:** TONY MCCONNELL/SCIENCE PHOTO LIBRARY; **p42:** Viktor Cap/Bigstock;
p52: EpicStockMedia/Bigstock; **p52:** JERRY SCHAD/SCIENCE PHOTO LIBRARY; **p52:**
Rafael Ben-Ari/Fotolia; **p54:** NASA; **p54:** PEKKA PARVIAINEN/SCIENCE PHOTO LIBRARY;
p55: AARON HAUPT/SCIENCE PHOTO LIBRARY; **p57:** ALEX CHERNEY, TERRASTRO.COM/
SCIENCE PHOTO LIBRARY; **p59:** LUKE DODD/SCIENCE PHOTO LIBRARY; **p60:** Scooter20/
Wikipedia; **p60:** NASA/JPL-CALTECH/MSSS/SCIENCE PHOTO LIBRARY; **p61:** Solar System
Exploration, NASA; **p61:** www.hubbesite.org; **p62:** Viktor Cap/Bigstock; **p62:** LARRY
LANDOLFI/SCIENCE PHOTO LIBRARY; **p63:** NASA; **p64:** ilker canikligil/Bigstock; **p65:**
LARRY W. KOEHN/SCIENCE PHOTO LIBRARY; **p67:** Georgios Kollidas/Bigstock; **p67:**
BABAK TAFRESHI, TWAN/SCIENCE PHOTO LIBRARY; **p69:** Abū Rayḥān al-Bīrūnī; **p70:**
Igor Chekalin/Bigstock; **p70:** DAVID A. HARDY, FUTURES: 50 YEARS IN SPACE/SCIENCE
PHOTO LIBRARY; **p71:** NASA/ESA/STSCI/S.BECKWITH, HUDF TEAM/ SCIENCE PHOTO
LIBRARY; **p71:** Konstantin Kulikov/Bigstock; **p74:** RIA NOVOSTI/SCIENCE PHOTO
LIBRARY; **p75:** Daniel Slocum/Bigstock; **p75:** VICTOR HABBICK VISIONS/SCIENCE
PHOTO LIBRARY; **p75:** formiktopus/Fotolia; **p75:** EcoShot/Bigstock; **p80:** Clive Rose/
Getty Images Sports/Getty Images; **p80:** Per-Anders Pettersson/Getty Images News/
Getty Images; **p80:** Yogesh More/Pantherstock; **p80:** Priscilla Gragg/Glow Images; **p82:**
Quinn Rooney/Getty Images Sports/Getty Images; **p82:** MistikaS/iStockphoto; **p83:**
Technotr/iStockphoto; **p86:** INDRANIL MUKHERJEE/AFP/Getty Images; **p86:** Jamie
McDonald - FIFA/Getty Images; **p88:** Uwe Kraft/Glow Images; **p90:** Holos/Stockbyte/
Getty Images; **p90:** Andrew Burke/Lonely Planet Images/Getty Images; **p90:** David Evans/
National Geographic/Getty Images; **p91:** Wayne Eastep/Photographer's Choice/Getty
Images; **p94:** Isaiah Love/Bigstock; **p94:** Classic Rock Magazine/Future/Getty Images;
p94: Rene Frederick/Photodisc/Getty Images; **p95:** Jake Lyell/Alamy; **p96:** Vladimir
Voronin/Fotolia; **p96:** Hit Delight/Fotolia; **p97:** Marc Broussely/Redfern/Getty Images;
p97: The Welfare & Medical Care/ Getty Images; **p98:** Steve Raymer/Asia Images/Glow
Images; **p99:** Sabphoto/Fotolia; **p99:** Fuse/ Getty Images; **p98:** Mark Lewis/Glow Images;
p103: Geostock/Photodisc/Getty Images; **p103:** ItinerantLens/Fotolia; **p103:** Martin
Ruegner; **p103:** Zandeuk/Bigstock; **p105:** Aflo Relax/Glow Images; **p105:** John Chapple/
Getty Images News/ Getty Images; **p106:** Cavan Images/Photonica/Getty Images; **p106:**
Superstock/Glow Images; **p110:** Triff/Shutterstock; **p110:** Mark Lewis/Glow Images;
p112:; Radius Images/Getty Images; **p112:** Datacraft - Sozaijiten/Alamy; **p112:** ©
belterz/iStockphoto; **p112:** Vladitto/Shutterstock; **p113:** Mike Flippo/Shutterstock;

p114: DETLEV VAN RAVENSWAAY/SCIENCE PHOTO LIBRARY; **p115:** Photodisc/ OUP
Picturebank; **p115:** ROBERT GENDLER/SCIENCE PHOTO LIBRARY; **p116:** David Gomez/
E+/Getty Images; **p116:** Jaggat/Bigstock; **p116:** India Today Group/Getty Images; **p116:**
Q2A Photobank; **p118:** Q2A Photobank; **p119:** Petr Josek Snr/REUTERS; **p120:** MARTYN
F. CHILLMAID/SCIENCE PHOTO LIBRARY; **p121:** Infomages/Bigstock; **p123:** JOHN GREIM/
SCIENCE PHOTO LIBRARY; **p124:** Luri/Bigstock; **p124:** GIPhotoStock/Photo Researchers/
Getty Images; **p125:** Everett Collection/Rex Features; **p125:** Gino Santa Maria/Bigstockl;
p125: Steve Allen/Brand X Pictures /Getty Images; **p128:** ROUF BHAT/AFP/Getty Images;
p128: Visage/Stockbyte/Getty Images; **p128:** Q2A Photobank; **p130:** Georgios Kollidas/
Shutterstock; **p130:** PHOTO RESEARCHERS/SCIENCE PHOTO LIBRARY; **p131:** ANDREW
LAMBERT PHOTOGRAPHY/SCIENCE PHOTO LIBRARY; **p133:** SCOTT CAMAZINE/SCIENCE
PHOTO LIBRARY; **p133:** POWER AND SYRED/SCIENCE PHOTO LIBRARY; **p133:** ARNO
MASSEE/SCIENCE PHOTO LIBRARY; **p134:** Visuals Unlimited/Getty Images; **p134:**
Gabe Palmer/CORBIS; **p134:** David Martyn/Bigstock; **p134:** CHARLES D. WINTERS/
SCIENCE PHOTO LIBRARY; **p134:** Jim Barber/Bigstock; **p134:** SCIENCE PHOTO LIBRARY;
p138: Virtual Photo/E+/Getty Images; **p138:** CORDELIA MOLLOY/SCIENCE PHOTO
LIBRARY; **p138:** CORDELIA MOLLOY/SCIENCE PHOTO LIBRARY; **p138:** CORDELIA
MOLLOY/SCIENCE PHOTO LIBRARY; **p139:** Sciencephotos/Alamy; **p140:** Andrea
Leone/Shutterstock; **p141:** SPENCER GRANT/SCIENCE PHOTO LIBRARY; **p142:** GI
PHOTOSTOCK/SCIENCE PHOTO LIBRARY; **p142:** GABRIELLE VOINOT/LOOK AT SCIENCES/
SCIENCE PHOTO LIBRARY; **p142:** ANDREW LAMBERT PHOTOGRAPHY/SCIENCE PHOTO
LIBRARY; **p144:** Ingersoll Rand Security Technologies ; **p145:** MEHAU KULYK/SCIENCE
PHOTO LIBRARY; **p145:** GEOFF TOMPKINSON/SCIENCE PHOTO LIBRARY; **p150:** JG
Photography/Alamy; **p150:** Ric Car/Bigstock; **p151:** MaZiKab/Bigstock; **p152:** Dave and
Les Jacobs/Blend RF/Glow Images; **p152:** Zero Creatives/Culture/Getty Images; **p152:**
Annapix/Dreamstime; **p152:** Creativefire/Dreamstime; **p153:** dozornaya/Bigstock;
p153: kavram/Shutterstock; **p153:** Ljupco Smokovski/Shutterstock; **p153:** luchschen/
Shutterstock; **p154:** Istvan Csak/Shutterstock; **p155:** ALEXIS ROSENFELD/SCIENCE
PHOTO LIBRARY; **p155:** Dan Barba/Stock Connection/Glow Images; **p157:** kadmy/
Bigstock; **p157:** Alvey & Towers Picture Library / Alamy; **p157:** Bloomberg/Getty Images;
p158: Nuttapong Wongcheronkit/Alamy; **p158:** Sergey Goruppa/Alamy; **p159:** Barry C.
Bishop/National Geographic/Getty Images; **p160:** Rainer W. Schlegelmilch/Getty Images;
p160: Stringer/AFP/Getty Images; **p166:** ChantalS/Fotolia; **p166:** Lenise Zerafa/Bigstock;
p167: Q2A Photobank; **p167:** LSU AgCenter; **p168:** LAWRENCE LAWRY/SCIENCE PHOTO
LIBRARY; **p168:** Art Director and Trip Photo Library; **p168:** BIOPHOTO ASSOCIATES/
Photo Researchers/Getty Images, MARK A. SCHNEIDER/SCIENCE PHOTO LIBRARY; **p169:**
Monty Rakusen/Cultura/Glow Images; **p171:** Jim Mills/depositphotos; **p171:** Eric Nathan/
Alamy; **p171:** MARTYN F. CHILLMAID/SCIENCE PHOTO LIBRARY; **p174:** Kim Karpeles
/ Alamy; **p176:** Greg ElMicrostock/Lonely Planet Images/Getty Images; **p177:** Rex
Features; **p206:** Q2A Photobank; **p208:** TONY MCCONNELL/SCIENCE PHOTO LIBRARY;
p210: mushakesa/Bigstock; **p210:** Q2A Photobank; **p212:** paul prescott/Fotolia; **p213:**
Riaan van den Berg/Bigstock; **p214:** NASA; **p216:** Jacek Fulawka/Fotolia; **p216:** koster62/
Fotolia; **p225:** Africa Studio/Fotolia; **p225:** Ludmila Smite/Fotolia; **p225:** Smileus/Fotolia;
p225: gchutka/iStockphoto

Artwork by: Q2A Media and Erwin Haya